plastics

basic industrial arts

First Edition

Lithographed in U.S.A.

DR. WILBUR R. MILLER
University of Missouri — Columbia
Columbia, Missouri

DR. GERALD L. STEELE
Ball State University
Muncie, Indiana

DR. MARION E. MADDOX
University of Arkansas
Fayetteville, Arkansas

LAVON B. SMITH
Fayetteville, Arkansas

McKNIGHT
Publishing Company
Bloomington, Illinois

Library of Congress
Card Catalog Number: 78-53391

SBN: 87345-788-9 Paperback
SBN: 87345-796-X Hardbound

Chapter 5 SECONDARY PROCESSES PL-72

Chapter 6 CAREERS IN THE PLASTICS INDUSTRY PL-84

Appendix PL-89

Chapter 1 INTRODUCTION TO PLASTICS

INTRODUCTION

For thousands of years, people have made useful things from moldable materials. These are materials that can be shaped by hand or in a mold. The two kinds of moldable materials are:

1. Natural (made by nature) and
2. Synthetic (made by humans).

Clay, glass, natural rubber, and shellac are all moldable **natural materials.** They have been in use for many years.

Synthetic materials are made by chemically changing natural materials. They have been developed to meet people's changing needs. **Plastics** is one of several groups of synthetic materials. See Fig. 1-1. Examples of other synthetic materials are foods, drugs, and grease. Plastics meet many needs formerly met by natural materials.

More than 100 years ago, a material was needed to replace ivory. To meet this need, John Wesley Hyatt invented cellulose nitrate in 1868. This was the first plastic. He used it to make billiard balls.

About 41 years later, in 1909, Dr. Leo H. Baekeland developed phenolic (phenol-formaldehyde) plastic. This plastic was called Bakelite.® It was soon molded into electrical insulators, bobbin ends, distributor caps, and other automobile ignition parts. The Eastman Kodak Company began using it for camera parts in 1914. John Hyatt even changed his billiard balls to Bakelite® phenolic resin in later years.

Many plastics have been invented since these first discoveries. Some of the most important ones, their uses, and dates of discovery are listed in Fig. 1-2. A more complete list can be found in the Appendix.

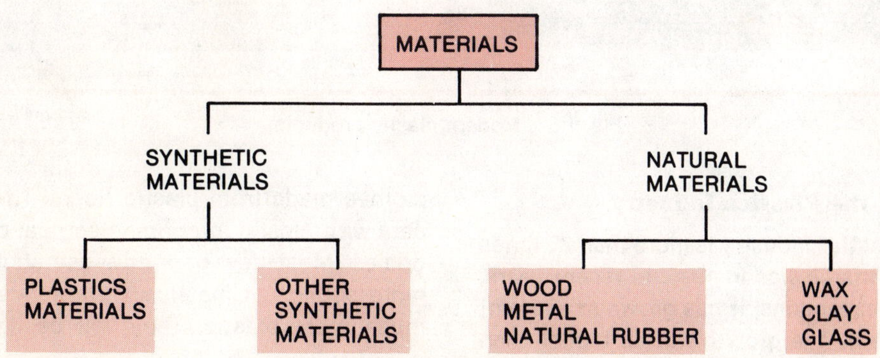

Fig. 1-1. Types of materials.

Date	Material	Main Uses
1868	Cellulose Nitrate (Celluloid®)	Little use today.
1909	Phenolic (Bakelite®)	Distributor caps, insulators.
1927	Polyvinyl Chloride (Vinyl or PVC)	Raincoats, auto upholstery fabric, shoes, notebook covers.
1936	Acrylic (Plexiglas®)	Airplane and snowmobile windshields, taillight lenses.
1938	Polystyrene (Styrene)	Squirt guns, toys, plastic car and airplane models.
1942	Polyethylene	Squeeze bottles, toys, bread wrappers, food containers, carpet fibers.
1942	Polyester	Fiberglass boat hulls, car bodies, coin embedments.
1943	Fluorocarbon (Teflon®)	Non-stick coatings, wire insulation for high-temperature uses.
1943	Silicone	Bathtub seal, rubber molds, space boots and gloves.
1947	Epoxy	High-strength glues, industrial molds and tooling.
1957	Polypropylene	Water ski tow ropes, carpet fibers.

Fig. 1-2. Historical development of selected plastics.

Modern Plastics Products

Fig. 1-3. Modern plastic products.

Growth of the Plastics Industry

The plastics industry is more than 20 times larger now than it was in 1945. In recent years, with some exceptions, it has grown at a rate of 15% per year. The growth rate is higher than any other industry.[1]

Today plastic products are found everywhere. People wear shirts, dresses, and other clothes made from plastic fibers. They ride in cars with plastic interiors, electrical parts, and bodies. Most new cars now use about 150 or more pounds of plastics. Future cars will use much more plastic. They will be lighter and

1. Recent statistics can be obtained from the Society of the Plastics Industry and Modern Plastics magazine.

more efficient. People use plastic telephones to talk with their friends. They buy detergents, milk, and other liquids in plastic containers. How many plastic items are used in your home?

Plastics is an important industry. It has already passed iron and steel in **cubic volume.** Plastic is lighter. One pound of plastic would fill the same space as 3 to 9 pounds of iron and steel. In other words, it takes fewer pounds of plastic than iron and steel to form equal cubic volume.

By 1983, more plastics will be produced than metals. Then, we will enter the age of plastics. According to United States Government reports, by the year 2000, plastics use will be three and one-half (3-1/2) times the volume of **all** metals.[2] In 1977, 32.6 billion pounds (about 14.8 million metric tons) of **plastics materials** were **produced** in the United States. About 227 billion pounds (103 metric tons) are predicted for the year 2000. The total **sales volume** was about $23 billion in 1970. About $634 billion is expected in the year 2000.[3] The amounts include all three parts of the industry. These are plastics materials, processors, and machinery manufacturers.

Plastics Education Foundation

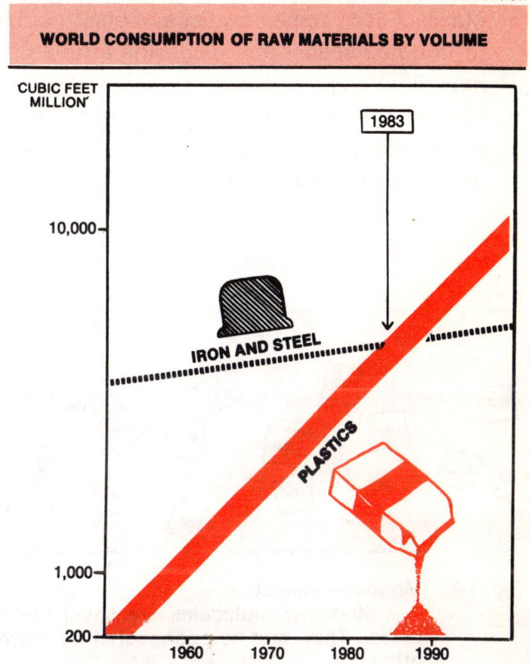

Fig. 1-4. Age of plastics.

Plastics Education Foundation

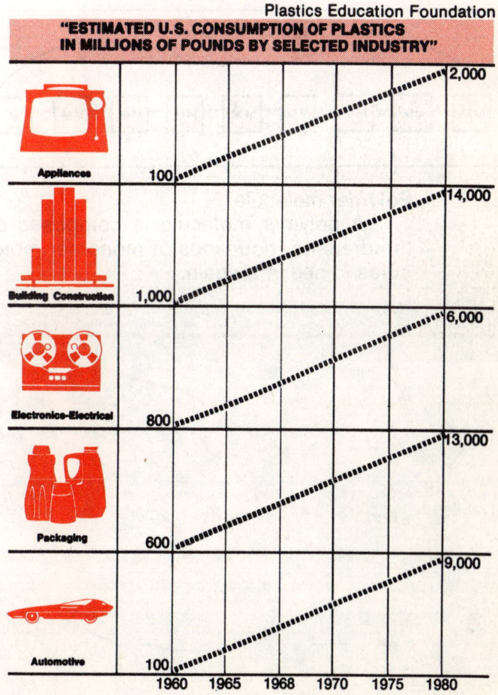

Fig. 1-5. Growth of plastics by selected industries.

STUDENT ACTIVITIES

1. Make a list of plastic products. Tell how each is used. Try to make your list different from those of your classmates.
2. Collect samples of plastic materials or products. Bring your collection to class. Try to find unusual uses for plastics.
3. Discuss the samples in class. Note the use of each plastic.

---

2. According to Government statistics quoted by J. Harry DuBois, President of Molecular Dialectrics, Clifton, New Jersey.
3. Taken from a report issued by the Society of the Plastics Industry, Inc. dated June, 1973.

4. Make a list of some **new uses** of plastics. Try to find newspaper and magazine articles on **new** plastics **products.**
5. List **all** the materials in an automobile (or some other large product). How many or what percent were plastics?

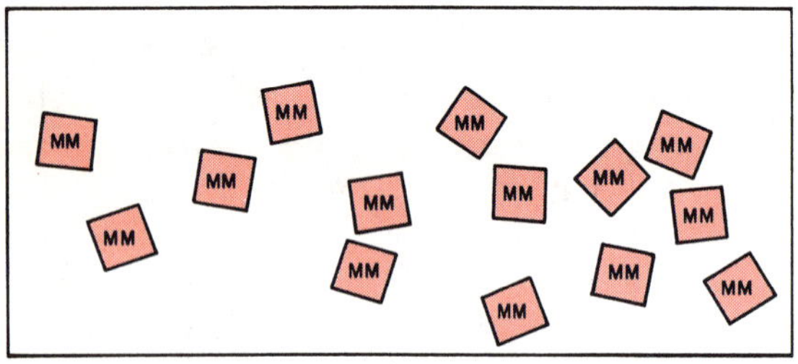

Fig. 1-6. Monomer molecules.
Monomer molecules are small molecules. They are not connected to each other.

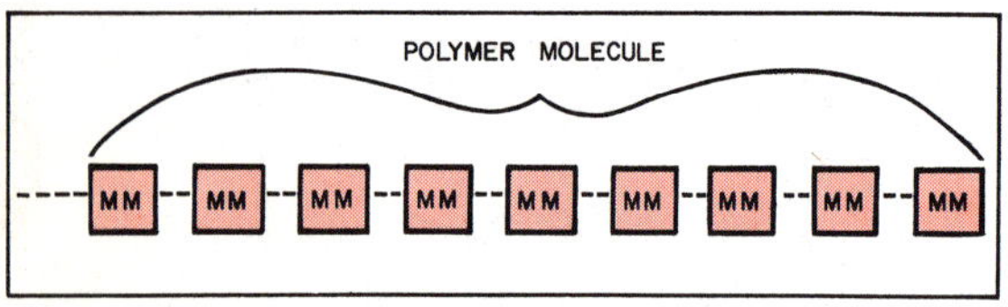

Fig. 1-7. Polymer molecule.
A polymer molecule is composed of hundreds to thousands of monomer molecules joined in a chain.

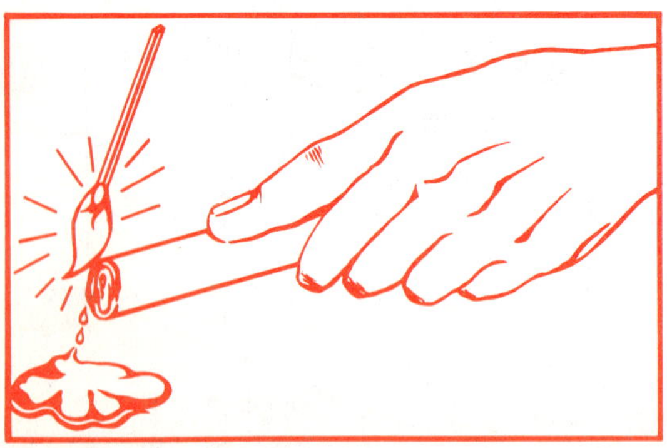

Fig. 1-8. Thermoplastics act like candle wax when heated or cooled.

DESCRIPTION AND CLASSIFICATION OF PLASTICS

The Composition of Plastics

Natural materials can be changed into new and different materials called **synthetics.** Plastics are synthetic materials. These are then made into useful products.

To make synthetic materials, chemists take atoms and molecules from natural materials. Some of these may be coal, oil, natural gas, air, water, plant fibers, and salt. Using heat, pressure, and catalysts, chemists rearrange these atoms and molecules to form plastics.

A substance in its simplest form is an element. Atoms are the smallest particles that keep the properties (qualities) of an element. Elements chemically joined together make a compound. A molecule is the smallest part of a compound that keeps the properties of a compound. It is made of two or more atoms chemically linked together. The size of a molecule depends on the number of atoms it has. Several small molecules may be combined to form a large molecule. The small molecules which are combined to form the large plastic (synthetic) molecules are called **monomer** (single unit) molecules. The large plastic molecules made from the monomer molecules are called **polymer** (many unit) molecules. These polymer (plastic) molecules are made up of hundreds to thousands of monomer molecules. They are linked together like a chain. Polymer chains are usually less than 1/10,000th of an inch (0.0025 mm) long, however.

Definition of Plastics

Plastics are synthetic materials made up of chains of molecules. During manufacture, plastics get soft and moldable under heat and pressure. Plastics are then hardened (solidified) into the molded shape.

The name "plastics" has come to mean a moldable or formable synthetic material. Natural materials such as glass, clay, and wax do not belong to this category. They are moldable but not synthetic.

Two Types of Plastics

All plastics are either:

1. Thermoplastic (heat softening) or
2. Thermosetting (heat curing).

Thermoplastics act like candle wax or ice. When heated, they make a **physical change** (change in state). A solid softens and then becomes a liquid. When cooled, they make the physical change from a liquid to a solid. The physical change takes place twice in thermoplastics during a molding cycle.

Thermoplastics may be softened and hardened many times. The plastics industry recycles (reuses) scrap thermoplastics. This conserves natural resources. It also prevents waste build-up. Community recycling drives collect thermoplastics for the same reasons. Thermoplastics may be degraded (destroyed) by too much heat, however.

Chemplex Company

Fig. 1-9. Polyethylene reactor system is shown in the foreground. In the background are silos where finished plastic resins are stored.

Thermosetting plastics are cured (set) into permanent shape during molding. Heat or heat and pressure cause a **chemical change** in them. This is a permanent change. It is similar to curing cement or baking a cake. Thermosetting plastics cannot be softened or recycled by reheating. They may be degraded by extremely high temperatures. They may also be cut or broken.

STUDENT ACTIVITIES

1. Using what you have learned, classify as thermoplastic or thermosetting plastic the samples you collected in Part I. Tag each sample. Save it for later use.
2. Make a list of scrap plastic materials from the home that can be recycled. Try to think of ways to reuse these materials.
3. Look for sources of scrap plastic materials in the community or around school. Some of these materials may be used in your classroom for processing.

Fig. 1-10. Thermosetting plastics act like concrete when set.

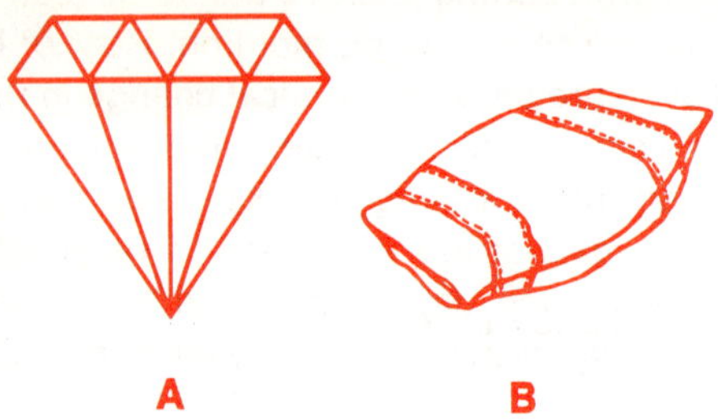

A. Hardness is like a diamond.
B. Softness is like a pillow.

Fig. 1-11. Hardness and Softness.

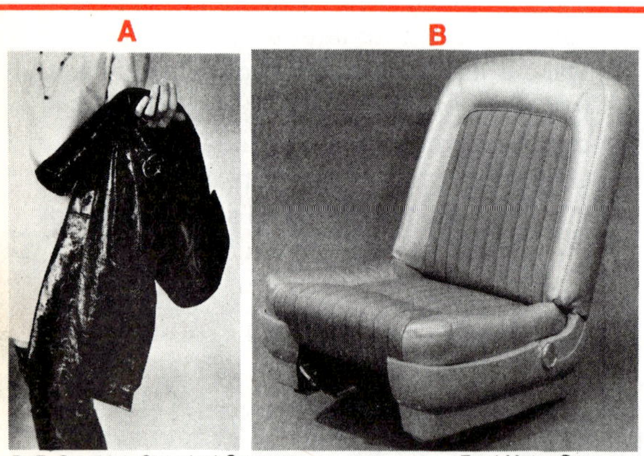

B. F. Goodrich Chemical Company — Ford Motor Company

A. Soft leather-like coat has a coating of thermoplastic polyurethane material.
B. Soft vinyl seat trim and nylon fabric cover this automobile seat. Flexible plastic foam is used for padding under the covering.

Fig. 1-12. Soft Plastic Products.

Fig. 1-13. Impact Strength.

PROPERTIES OF PLASTICS

Plastics have many outstanding **properties** or qualities. These properties change greatly from family to family of plastics. They also change widely within each family. A **family of plastics** is a group of plastics which is chemically the same. Each family of plastics has a name. Some of these are polyethylene, polypropylene, acrylic, and phenolic.

Hardness and Softness

Plastics can be made hard and stiff or soft and flexible. Usually, lightweight plastics are soft and flexible. Often, heavy plastics are hard and stiff. Some hard, stiff plastics break easily. Others do not.

Phenolic plastic is hard and stiff. Phenolic is used for auto distributor caps, washing machine agitators, and pot handles.

Another hard, stiff plastic is polyester. It is used with fiberglass to make boat hulls, car bodies, and fishing rods. The glass fibers give strength. The polyester plastic makes them rigid. Polyester resins may also be used for clear castings.

One of the softest plastics is low density polyethylene. It is used for detergent bottles, bread wrappers, dry cleaners' bags, and toys. Another soft, flexible plastic is polypropylene. It is often used for rope. Carpeting for indoor or outdoor use may be made of polypropylene fibers. Polypropylene is also used for one-piece plastic hinges.

Impact Strength

Plastic products should be able to withstand being dropped without breaking. This is called **impact strength.**

Most thermoplastics have acceptable impact strength. Low density polyethylene has very good impact strength. It is also quite flexible. Most detergent bottles are made from polyethylene.

Polyvinyl chloride (vinyl) has very good impact strength also. It is used for notebook covers, furniture, and automobile upholstery. Polycarbonate plastic has one of the highest impact strengths of all plastics. Ionomer plastic also has high-impact strength.

Many common thermoplastics are mixed with glass fibers to give them better impact strength. These are called glass-filled thermoplastics. An example is glass-filled polypropylene. It is used for automobile inner fenders and fan covers.

Impact strength is also important in thermosetting plastics. A combination of epoxy plastic and fiberglass is excellent. It probably gives the best impact strength for its weight of all thermosetting plastics. Epoxy/glass materials are used for space rocket parts. These parts include fuel tanks, covers, and heat shields. Polyester resins with fiberglass have high-impact strength. They are used for boats, car bodies, and snowmobile hoods. Power tool handles and rocket engine parts are often made of glass-filled phenolics.

Electrical Insulation

Plastics, as originally made, are good electrical insulators. They will not carry an electrical current. Some plastics are better electrical insulators than others. Also, some plastics may be **made** conductive (able to carry electricity). This is done by adding metallic fillers to the resin.

An important thermoplastic electrical insulator is low density polyethylene. It has high insulation value and is flexible. It is excellent to use for electrical wires. These properties made television and radar cables possible. Fluorocarbon insulation is also excellent for electrical wires and cables. It can stand very high temperatures for a long time. One use is for wire insulation in large electrical transformers. Fluorocarbon is more expensive than polyethylene. Polyvinyl chloride is used for low-frequency electrical insulation. It will stand more heat before melting than polyethylene.

A rigid thermosetting plastic, phenolic, has high electrical insulation value. It is used for rigid parts. Some of these are automobile distributor caps, rotors, and electric coffeepot bases.

Weight

Plastics are considered lightweight solid materials. They are lighter than most metals. Some plastics are light enough to float on water. Two of these are polypropylene (featherweight plastic) and polyethylene (squeeze bottle plastic). Polystyrene and polyvinyl chloride are heavier than water and will sink. Most thermosetting plastics are heavier than water. These include phenolics, polyesters, and epoxies. They will sink.

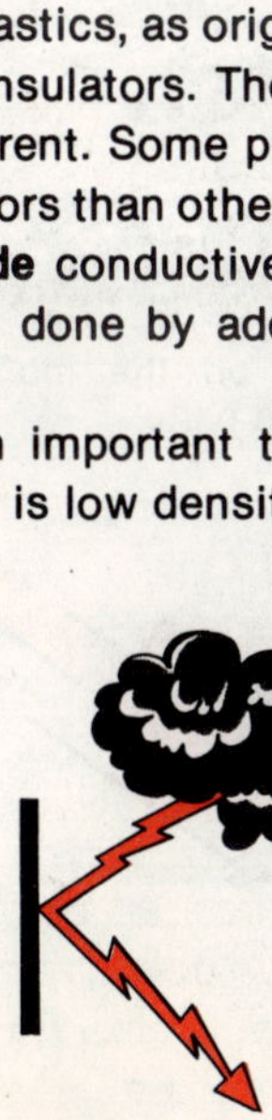

Fig. 1-14. Electrical insulation.

Arco/Polymers, Inc.

Fig. 1-15. Lightweight.

Arco/Polymers, Inc.

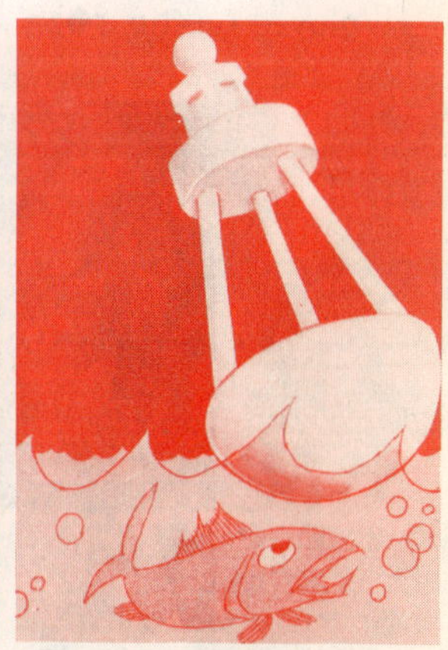

Fig. 1-16. Lightweight plastics float on water.

Heat Insulation

Plastics are good heat insulators. Sometimes they are foamed or mixed with fillers like asbestos. This makes them better heat insulators.

Rigid polyurethane foam is considered the best heat insulating plastic. Foamed polystyrene is also very good. It is used for picnic coolers and ice buckets (expandable polystyrene bead foam). Throwaway hot drink cups and large building insulation panels are made from polystyrene bead foam.

Polystyrene foam (Styrofoam®) is formed into boards. The boards are used for building insulation. Styrofoam is also used for insulating refrigerators and freezers.

Sometimes heat insulators must be strong and rigid. Good rigid heat insulators can be made from phenolics mixed with asbestos. This material is used to make pot and pan handles and electric coffeepot bases.

Temperature Resistance

Materials are needed which will take extreme temperatures, both hot and cold. Silicone is used where a flexible material is needed. It can be used for gaskets, glues, sealers, and molds. Silicone is a thermosetting plastic. It will stand up to 900° F (482° C) for a few minutes. Astronauts wore silicone boots and gloves on the moon.

Arco/Polymers, Inc.

Fig. 1-17. Foamed polystyrene insulated cup.

Phenolic plastics have good high-temperature resistance. They are often used for electric frying pan and iron handles.

Fluorocarbon plastics will stand extremes of temperature. These may range from − 400° F to + 500° F (− 240° C to + 260° C). Fluorocarbons are thermoplastics. They are used for wire insulation. They do well where temperature ranges from very hot to very cold. One such place is in a space rocket.

E. I. DuPont De Nemours & Company

Fig. 1-18. Space suits worn on the moon have Teflon® fluorocarbon parts.

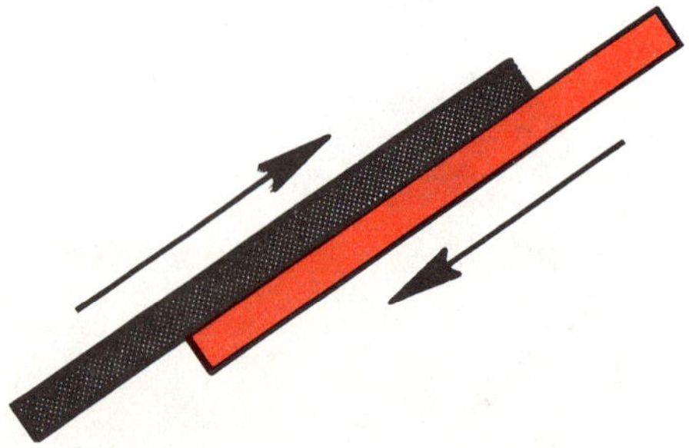

Fig. 1-19. Friction.

Friction

Several plastics are low in friction (slippery). Fluorocarbon plastics (Teflon®) are among the best. Most things will not stick to them. Frying pans, electric irons, saw blades, and other products are coated with Teflon®.

Polyethylene has the same nonstick quality. It melts at a much lower temperature than fluorocarbon (400° F or 205° C less). It cannot be used in places with high temperatures. It can be used as a mold for casting polyester plastic resins.

Dow Corning Corp.

Fig. 1-20. Astronaut uses silicone rubber hoses during a walk in space. The hoses are flexible and resist extreme temperatures.

Arco/Polymers, Inc.

Fig. 1-21. Cushioning.

Elasticity

Elastomer plastics are stretchy. They are like rubber bands. Elastomers can be stretched to twice their length at 72° F (22° C). Then they return to their original shape. Silicone is a good example. It is used as a flexible sealer and glue. It may be used for bathtubs, airplanes, and space rockets. Metal and plastic parts are cast in silicone molds.

Transparency

Transparency (clearness) varies with plastics. Some are as **clear** as glass. **Opaque** plastics cannot be seen through. **Translucent** plastics can be partly seen through. Shapes can be seen, but not details.

Natural polystyrene is crystal clear. It is often colored. Clear polystyrene is used for food containers. It is inexpensive, odorless, and tasteless. The food inside can be seen.

Acrylic plastic (Plexiglas® or Lucite®) is used for many clear, outdoor products. It stands the weather well. It is stronger than polystyrene. Airplane windshields, signs, and automobile taillight lenses may be made of acrylic plastic. Low-cost camera lenses and contact lenses may be molded from acrylic plastic.

Clear polycarbonate has high impact strength. It is often used for taillight lenses.

Cushioning

Cushioning is how well a material absorbs shocks or impacts. Cushioning materials protect products from damage during shipment. Cushioning materials are used in cars to protect people. For example, they may be used on the dashboard or steering wheel. They are also used in chairs and mattresses. Most cushioning materials are made from plastic foams. They may be either rigid or flexible. Expandable polystyrene bead foam is a cushioning material.

Kohler - General Plastics Division

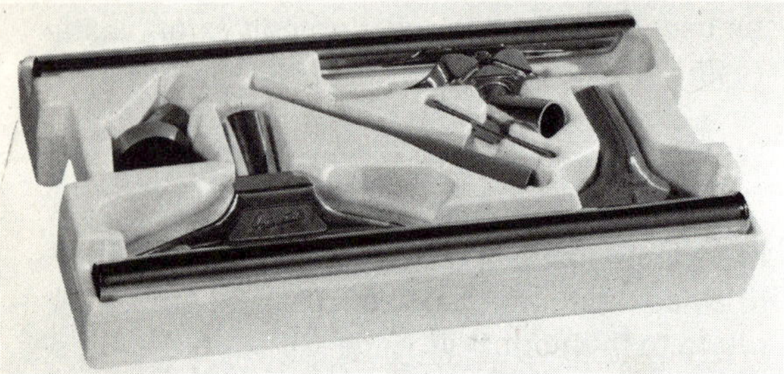

Fig. 1-22. Products are packed in bead foam plastics with cushioning characteristics.

Other examples are polyurethane (either flexible or rigid) and polyethylene foam.

Plastics Are Many Things

Any material has many properties (qualities). The chemist matches the needs of a product or use for it with the properties of a material. Sometimes present materials do not meet the needs. A new material must be built. The chemist arranges atoms and molecules to suit the needs. The new materials are synthetics. Plastics are synthetics. Properties of plastics are determined by the chemical makeup. The properties of many plastics can be compared in Fig. 1-23.

STUDENT ACTIVITIES

1. Discuss the main qualities of the sample plastics collected for class.
2. Discuss why each product needs the qualities of the plastic from which it is made.
3. Try to find ways in which these products can be improved by changing the plastic from which they are made. What qualities should be improved and why?

	Property → / Family ↓	Hardness	Impact Strength	Electrical Insulation	Weight	Heat Insulation	Temperature Resistance	Friction	Elasticity	Transparency
THERMOPLASTICS	ACRYLIC (Lens)	H	M	M	M	M	M	M	M	H
THERMOPLASTICS	FLUOROCARBON (Nonstick)	H	H	H	VH	VH	VH	VL	M	L
THERMOPLASTICS	POLYETHYLENE, HI DENSITY (Milk bottle)	H	L	H	L	H	L	L	L	M
THERMOPLASTICS	POLYETHYLENE, LO DENSITY (Squeeze bottle)	L	VH	VH	L	M	VL	VL	H	M
THERMOPLASTICS	POLYPROPYLENE (Featherweight)	M	M	H	L	M	M	L	M	L
THERMOPLASTICS	POLYSTYRENE GEN. PURPOSE (Squirt Gun)	H	L	M	L	L*	VL	H	L	VH
THERMOPLASTICS	POLYVINYL CHLORIDE (Imitation leather)	M	H	M	M	L	VL	M	M	M
THERMOSETTING	EPOXY (Glue)	H	H	H	L to M	M	H	M	L	H
THERMOSETTING	PHENOLIC (Distributor cap)	VH	H	VH	M	H	H	H	VL	None
THERMOSETTING	POLYESTER (Boat)	H	L	H	M	M	M	L to M	VL	H
THERMOSETTING	SILICONE (Bathtub seal)	L to M	H	H	M	VH	VH	L	H	L

VL = Very Low
L = Low
M = Medium
H = High
VH = Very High

Fig. 1-23. Qualities (properties) of selected plastics as compared to other plastics. All materials in natural unaltered state.

4. Consider a plastic product that is to be used only once. Why was the plastic chosen for one-time use?

5. Play the game **Plastic.** Study Fig. 1-24. This game is played much like Bingo. Prepare your own game sheet. **Do not** write in this book. First, study the five columns of words near the top of Fig. 1-24. On a sheet of paper write the word "plastic" in separated letters as shown. Place them as headings for five columns. Select five words from those listed under the letter P. Write them in a column under the P. Select five words from those listed under L. Write them in a column under L. Continue in the same way for each column. Next, rule off a form like the one shown. Number the five rows of spaces vertically. Letter the five columns of spaces horizontally. Write "free" in the center space. Your teacher will call out a word from those listed in Fig. 1-24. You will be told the space where it may be written on your form. The teacher may say "P — 1" or "L — 2." Your teacher will then ask a student what the word means. If the student's answer is correct, all those who listed that word on their paper can write it in the given space. If the answer is wrong, no one can use the word until someone correctly defines it. This continues until a student has five words in a row. The five words may be counted horizontally, vertically, or diagonally. Remember, the word can only be written in the place on your form that the teacher gives.

6. Design new plastic game cards.

1	MATERIALS	MOLDED	FORMED	PRESSURE	MOLDING
2	PRODUCTS	SYNTHETIC	NATURAL	ATOM	MOLECULE
3	MONOMER	POLYMER	PLASTIC	THERMOPLASTIC	THERMOSETTING
4	PROPERTIES	PLASTIC FAMILY	CATALYST	HOMOPOLYMER	COPOLYMER
5	TERPOLYMER	CHEMIST	PRODUCED	CELLULOID	HYATT
6	BAEKELAND	BAKELITE	PHENOLIC	POLYVINYL CHLORIDE	ACRYLIC
7	POLYETHYLENE	CURE	HARDNESS	IMPACT	INSULATION
8	RESISTANCE	ELASTICITY	TRANSPARENCY	OPAQUE	TRANSLUCENT

	P	L	A	S	TIC
1					
2					
3			FREE		
4					
5					

Fig. 1-24. Plastics game.

MOLDING

Chapter 2

During molding processes, plastic material may be compressed in a mold or die. Or it may be forced into or through a mold or die. Plastic materials are formed or finished under pressure. Molding includes injection molding, extrusion, blow molding, compression molding and transfer molding.

Almost all plastics go through one or more of the molding processes. Equipment and molds are more expensive than for other methods. However, these processes are economical for long production runs at high production rates. Parts produced may be very small or as large as 100 lbs. (45 kg). More plastics are produced by molding than by all other processes combined.

INJECTION MOLDING

Process Description

Injection molding is a widely used process. Hot, soft plastic is injected (pushed or forced) into a closed mold. The material is cured or hardened in the mold. Both thermoplastic and thermosetting plastics can be injection molded.

Thermoplastic injection molding is the most widely used. The thermoplastic material is

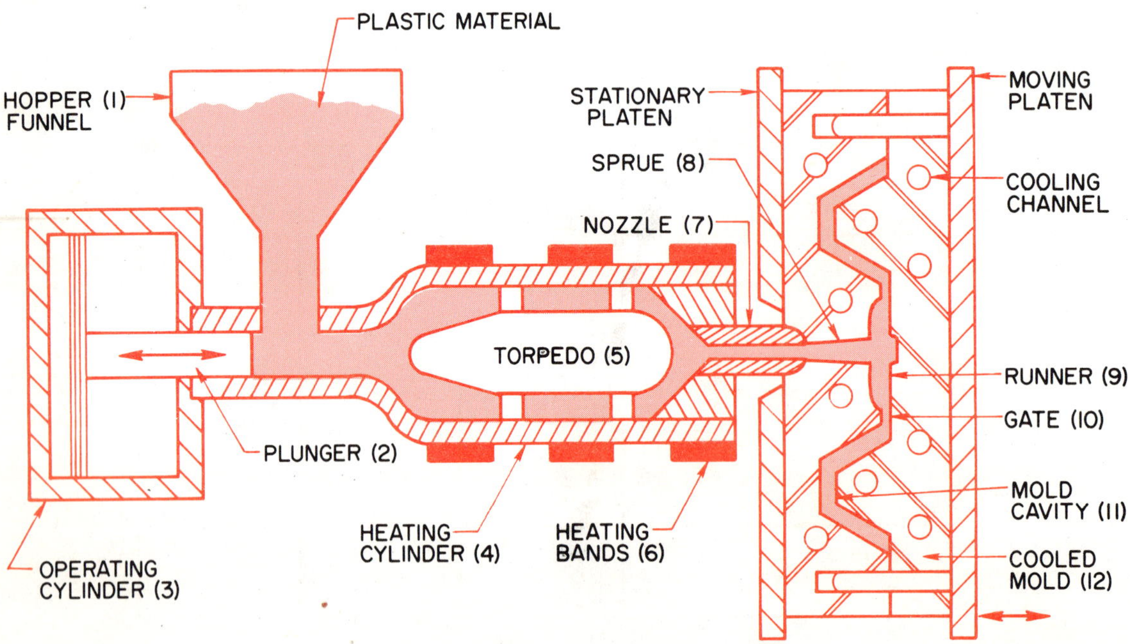

Fig. 2-1. Horizontal plunger injection molder.

hardened by **cooling** it in the mold. This is a **physical change.**

Thermosetting injection molding is a fairly new process. It needs special injection unit parts. Thermosetting plastic is cured by **heating** it in the mold. This is a **chemical change.**

There are two types of injection molding machines. These are: (1) plunger injection and (2) screw injection. All injection molders have two sections: (1) an injection section and (2) a mold operating section.

Plunger Injection Molding

Plunger injection molding (sometimes called ram injection) is an older type of injection molding than screw injection molding. Plunger injection molders mold all types of thermoplastics. Laboratory-size injection molders for schools are often plunger machines. They should be used to mold only thermoplastics.

To follow the operation of the plunger injection molder, study Fig. 2-1. Thermoplastic molding material is placed in the **hopper funnel** (1). It is dropped by gravity into the feed throat. The **plunger** (2) pushes it forward. The **operating cylinder** (3) moves the plunger back and forth. The plastic is moved into the **heating cylinder** (4). It is pushed against the walls by the **torpedo** (5). The cylinder is usually heated with electric **heating bands** (6). The plastic is heated as it passes between the torpedo and the heating cylinder walls. The hot plastic then goes through the **nozzle** (7). The nozzle fits tightly against the **sprue** (8). The sprue carries the plastic into the mold. The hot plastic then flows through the **runners** (9) and **gate** (10) into the **mold cavity** (11). Refer to Figs. 2-1 and 2-6.

The plastic cools and hardens in the mold under pressure. The mold is then opened and the part is pushed out. The product of one injection molding cycle is called a **shot.** One shot may have one or more parts. It includes the part and also the sprue, runner(s), and gate(s).

Screw Injection Molding

Screw injection molders are similar to plunger injection molders. The main difference is in the heating cylinder. See Fig. 2-2. The plastic is fed by gravity from the **hopper funnel** (1) into the **cylinder** (2). A turning **screw** (3) moves the plastic forward. This causes fric-

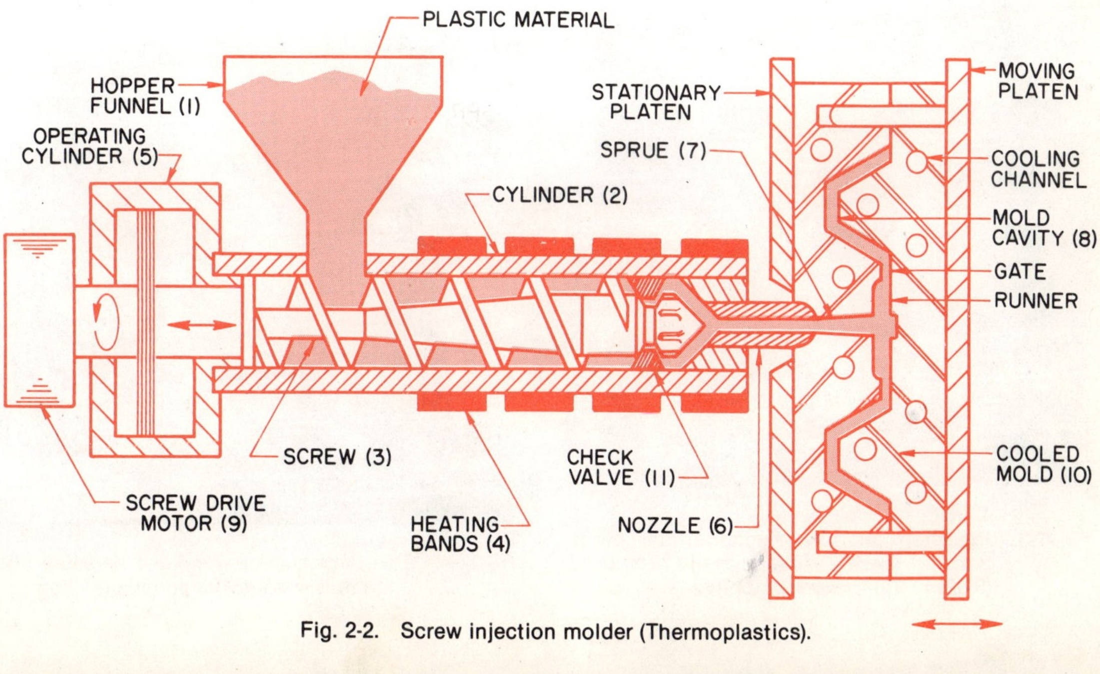

Fig. 2-2. Screw injection molder (Thermoplastics).

tional heat in the plastic. Electric **heating bands** (4) heat the outside of the cylinder, also. The heat melts the moving plastic. The plastic moving forward pushes the screw backward. Plastic builds up ahead of the screw. The screw stops turning when enough plastic is built up. The screw is then pushed forward by the **operating cylinder** (5). This injects the plastic. A **check valve** (11) is on the front of the screw. It keeps the plastic from sliding backward during injection. The plastic flows through the **nozzle** (6) and **sprue** (7). It then enters the **mold cavity** (8). There the plastic cools. Meanwhile, more plastic is heated by the turning screw. The mold opens after the screw stops turning. The part is then pushed out from the mold.

The screw injection molder is the newer kind of machine. It has almost replaced the plunger injection molder in industry. The plastic is melted with less chance of chemical damage. The screw injection molder also mixes the plastics and colors better.

Importance, Advantages, and Disadvantages of Injection Molding

Injection molding is considered the most important plastics process because it produces more types of products than any other. It is probably the most widely used plastics process. More injection molding machines are in use today than any other type of plastics processing machine. Over 27% of the plastics used in the U.S. are injection molded. By 1980, an increase to about 32% is expected.

Injection molding is used to produce long production runs of parts. They are produced at a very fast rate. Parts of various sizes can be made. Some are small. Note the size of the telephone parts shown in Fig. 2-3. Some are as large as the part shown in Fig. 2-4. A wide range of thermoplastics may be used. Also, many thermosetting plastics are screw injection molded. Mold costs for production injection molding are high. Strong, expensive machines are necessary. Molds and machines must be built to take high molding pressures.

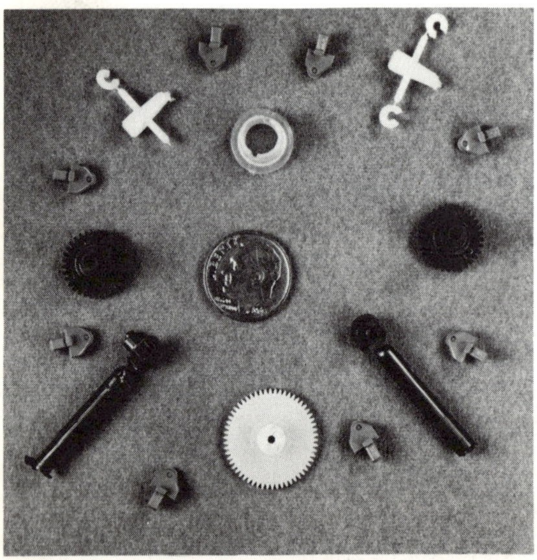

Fig. 2-3. Precision molded telephone dial parts. Some of these parts require fits of ± .001″ (0.0254 mm) to work properly.

Fig. 2-4. This large injection molded part is a filter plate that weighs 100 pounds (45 kg).

Industrial Equipment

Industrial injection molding equipment produces shots of different sizes. A **shot** is the amount of general purpose polystyrene plastic that can be injected in one stroke (shot) of the injection plunger or screw. The range of sizes is wide. Some equipment produces shots of about 1 ounce (28 g). Some equipment produces shots of 800 ounces (23 kg) or larger.

Materials Used in Industry

Almost all thermoplastics may be injection molded. Common thermoplastics used are acrylic, cellulose acetate, polyethylene, polystyrene, and polypropylene. Some high-temperature and special thermoplastics may need special injection molding machines. These include fluorocarbon (Teflon® or TFE), nylon, and polyphenylene oxide (PPO).

Many thermosetting plastics are being injection molded. Specially designed screw injection molding machines are used. Examples of injection molded thermoset plastics are phenolic, melamine, urea, rubber, and thermoset polyester.

Time, Temperature, and Pressure Relationship

Time, temperature, and **pressure** have a very close relationship when plastics are molded. They are very important in all plastics processes. The relationship is like a triangle, see Fig. 2-7. Changing one affects the others. An **increase** in one usually means a **decrease** in another. A **decrease** in one usually means an **increase** in another.

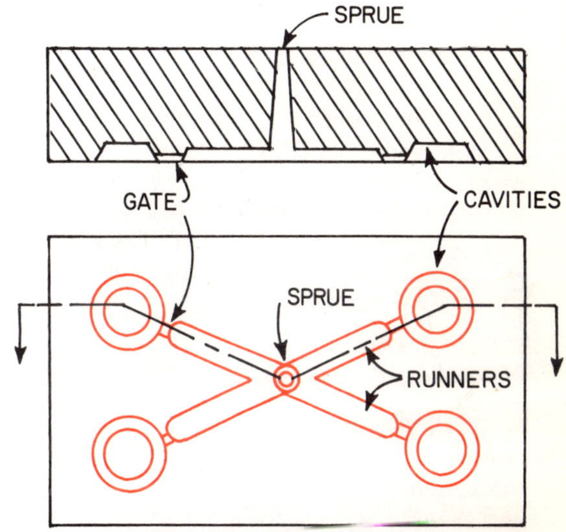

Fig. 2-6. A typical sprue, runner, and gate system for cavity side (rear) injection. Note how the system is balanced. All of the runners are the same length and shape. All the cavities are the same distance from the sprue.

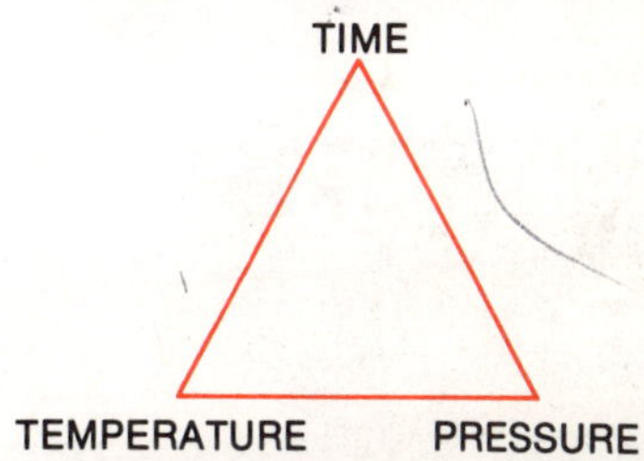

Fig. 2-7. Time, temperature, and pressure relationship.

National Automatic Tool Mfg. Company

Fig. 2-5. Large single-stage screw injection molder with 2700 ton (2449 Mt) clamp and 300 ounce (8505 g) styrene injection capacity. Note worker's size as related to machine.

Checklist of Materials and Equipment Needed

The following materials and equipment are needed for injection molding:

1. Injection molder.
2. Injection mold(s).
3. Injection molding plastic (type depends on mold and product).
4. Mold release — silicone or fluorocarbon.
5. Wrenches for adjusting the machine (if needed).
6. A small wood dowel about 8″ to 12″ (20 to 30 cm) long.

Operating Temperatures

Cylinder temperatures for plunger machines are usually about 100° F (55° C) above the softening point of the plastic used. Fig. 2-18 will serve as a guide for molding temperatures of several common injection molding plastics. Start on the low end of the range. Slowly increase the cylinder temperature. Stop when the plastic flows easily from the nozzle. Keep the molding cycle the same. This will keep the time the plastic is melted (residency time) the same. Keeping the residency time the same will help keep the molded parts identical.

Injection Molding Operation Sequence

There are five steps in the injection molding process.

1. Prepare the machine for operation.
2. Melt the plastic **thoroughly** in the heating cylinder.
3. Inject the right amount of plastic into the mold. Use enough injection pressure to make a good surface finish on the part. The part should be free of weak welds, flow marks, sinks, and other defects.
4. Cool the plastic fast enough to harden it.
5. Open the mold for removal of the part. Close the mold for the next molding cycle.

The molding sequence is shown in the following illustrations. **Look at the machine instruction manual for exact operation information.**

1A

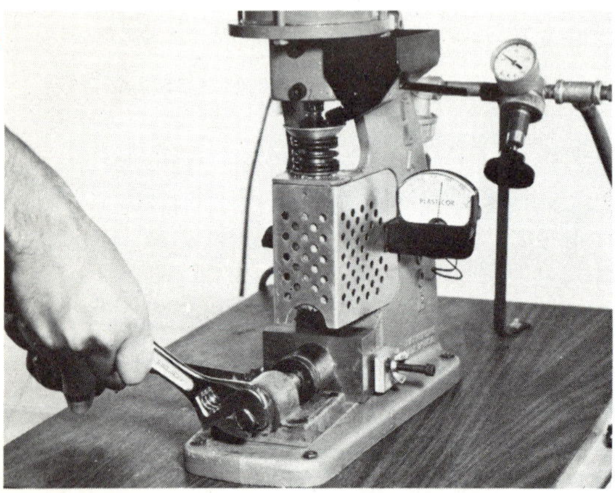

Fig. 2-8. Adjust the machine to fit the mold.

1B

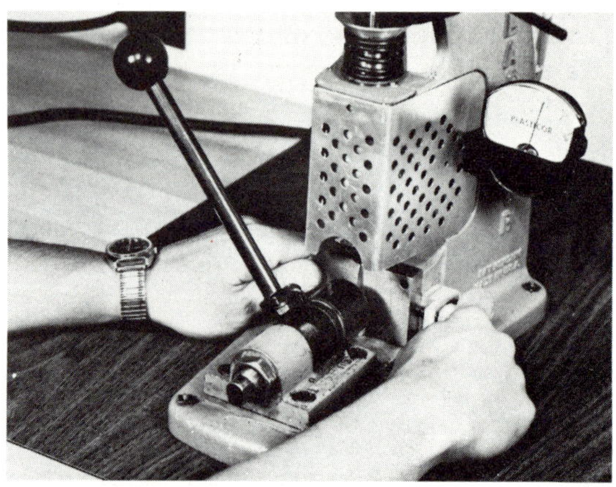

Fig. 2-9. Adjust the mold centering screw.

1A Set the clamp pressure (if needed) for a firm over-the-center snap.

1B Plug in the electrical cord. Turn on the machine and warm it up to operating temperature. Check the plastic chart for various molding temperatures. (See Fig. 2-18.)

Fig. 2-10. Place material in the hopper.

2A

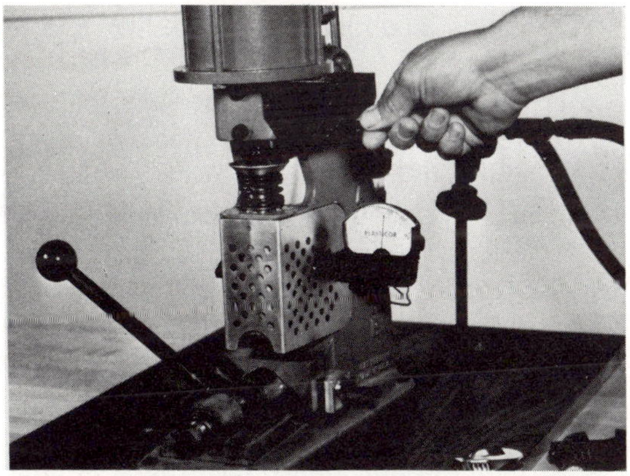

Fig. 2-11. Fill the injection cylinder.

2A Allow the material to preheat in the cylinder. **Caution! Do not overfill the cylinder.** No material should stay in the cylinder **funnel** at **any** time.

2B

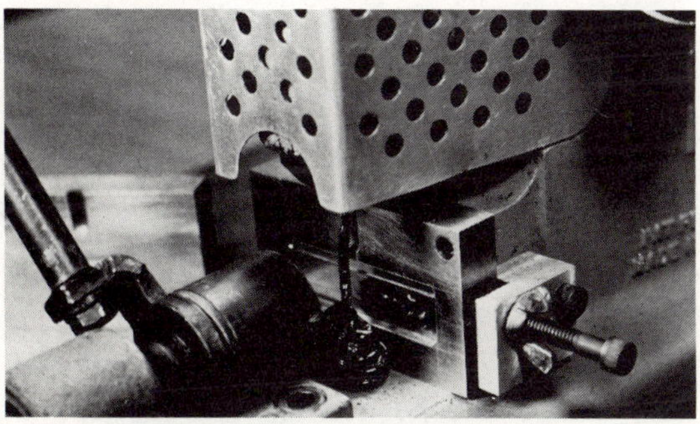

Fig. 2-12. Purge the injection cylinder.

2B Connect the air line if it is an air-operated press. Purge the cylinder before injection. Material must be at operating temperature (plasticized).

2C

Fig. 2-13. Spray on mold release.

2C Put a little mold release on the mold. Apply only as needed. **Do not put on for each shot.**

3A

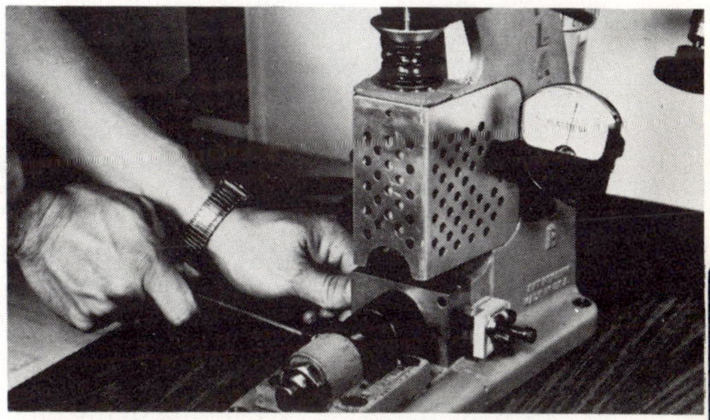

Fig. 2-14. Clamp the mold into the machine.

3A **Check** sprue and nozzle alignment.

3B

Fig. 2-15. Inject the material into the mold.

3B (a) Operate manual injection lever down, OR
(b) Operate 4-way air valve forward or downward (injection direction).

4A

Fig. 2-16. Operate the injection lever.

4A (a) Hold pressure on the material for a short time (several seconds). This is known as **dwell time or injection forward time.**

(b) Retract the plunger.
Operate the manual injection lever up,
OR
Operate the 4-way air valve back or upward (retract direction).

(c) Refill the injection cylinder to the proper height.

5
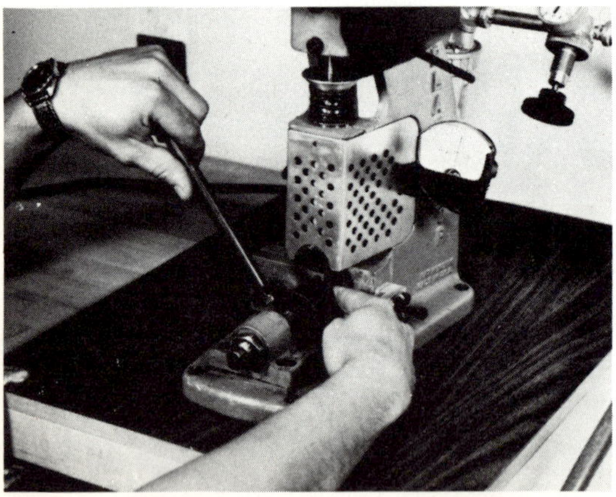

Fig. 2-17. Open the mold clamping device.
Remove the part from the mold,
OR
5 Open the clamping device and allow the ejector pins to push out the part (semiautomatic).

STUDENT ACTIVITIES

1. Process a variety of materials in each available mold. Use only polyethylene for undercut molds, such as bottle caps.
2. Experiment with plastics materials.
   a. Blend two or more colors of the same material. Try to get a variety of color effects.
   b. Blend two different thermoplastic materials together and mold. What differences are noted in the products produced? **Caution:** Do not use undercut molds, such as bottle caps, for this experiment.
   c. Blend two densities of the same material. This will change the density of the material. Note how the blend averages to a density between the two parent materials.
3. Experiment with the injection molding process. **Change only one variable at a time.** Make several shots at each setting you use.
   a. Vary the injection molding **temperature.** Record and analyze the results.
   b. Vary the injection forward **time** in injection molding. Record and analyze the results.
   c. Vary the injection molding **pressure.** Record and analyze the results.
4. Experiment with an educational toy injection molder. Compare the process and products with the laboratory injection molder.

EXTRUSION

Process Description

In the extrusion process, plastic materials are heated and pushed forward by a turning screw. The materials are forced through an opening called a **die.** The die shapes the plastic. Long, continuous shapes are made by the extrusion process. Some examples are garden hoses, soda straws, and plastic rods. Plastic and rubber wire insulation is applied by extrusion. Thin plastic film is made by this process. It is used for items such as sandwich and dry cleaning bags. Paper and cardboard are often coated with plastic by extrusion. Most blow molders use an extruder to heat and melt the plastic. Many thermoplastics used for injection molding are first extruded into small rods. Then they are chopped into plastic pellets. Thermoplastics, thermosetting plastics, rubber, and food products can be extruded.

Refer to Fig. 2-19 to follow the extrusion process. Plastic pellets are loaded into the **hopper** (1) of the extruder. They are dropped by gravity into the **barrel** (3). The turning **screw** (2) moves the plastic forward. The screw turns all the time that the extruder is working. It does not start and stop like the injection molder screw. As the material moves forward in the barrel (3), it is heated. The heat comes from two sources. These are outside **heating bands** (4) and inside frictional heat. Thermostats control the heat from the outside heating bands. Inside the barrel, the screw presses and rubs the plastic particles together. This causes **frictional heat.** Seventy-five percent (75%) or more of the heat added to the plastic in extrusion is from friction.

Plastic Material (Family Name)	Melting Point	Molding Cylinder Temperature Range
ABS (Medium to High-Impact)	325° F (163° C)	380° to 525° F (193° to 274° C)
Acrylic	374° F (190° C)	400° to 520° F (204° to 271° C)
Polyethylene, Low Density	221° F (105° C)	225° to 325° F (107° to 163° C)
Polyethylene, High Density	248° F (120° C)	275° to 500° F (135° to 260° C)
Polypropylene	334° F (168° C)	400° to 550° F (204° to 288° C)
Polystyrene, General Purpose	374° F (190° C)	325° to 500° F (163° to 260° C)
Polystyrene, High Impact	374° F (190° C)	350° to 600° F (177° to 315° C)

Fig. 2-18. Common injection molding temperatures for thermoplastics.

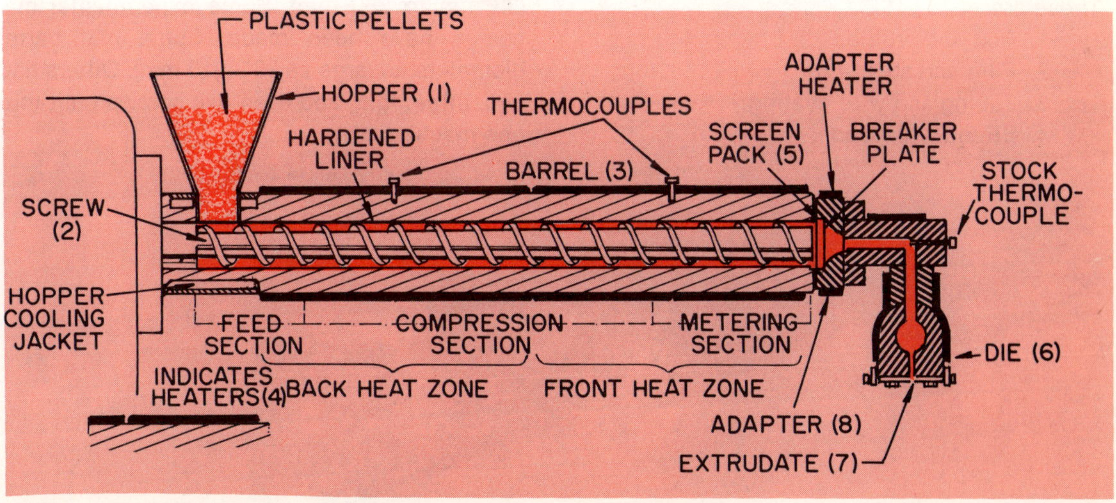

Fig. 2-19. Extruder with flat sheet die.

The melted plastic passes through a **screen pack** and **breaker plate** (5). This traps impurities in the plastic. It also straightens the flow. Refer again to Fig. 2-19. The screen pack keeps the die from clogging.

The plastic enters the **die** (6). It takes on the shape of the die opening. The plastic that comes through the die is called **extrudate** (7). The **die adapter section** (8) and die are also heated with heating bands. The heating bands are usually electrically powered.

Extrudate is taken away and cooled by a take-off system. As it is cooled, it solidifies. This is a **physical change.** This system can be regulated. It must take the plastic away from the extruder at the right speed. The speed of both the take-off system and extruder are changeable.

The extruder has at least three heating zones. These are:

1. Rear barrel,
2. Front barrel, and
3. Die.

Extruder Dies, Cooling, and Take-off Equipment

A number of **die types** are used for plastics. They may be divided into four groups. These are:

1. Rod and profile,
2. Film and sheeting,
3. Tubing and wire coating,
4. Strand and ribbon.

Three types of **cooling** are used for plastic extrudate. These are:

1. Forced air,
2. Water spray and water quench (dip),
3. Chill rolls.

Extrusion **take-off equipment** may be divided into two types:

1. Pinch rolls,
2. Caterpillar tread units.

Each extrusion system is a combination of an extruder, a die, coolers and take-off units.

Importance, Advantages, and Disadvantages of Extrusion

More plastic materials are processed by extrusion yearly than any other process. About 33% of all plastics materials are now extruded. That figure does not include blow molding. When blow molding is added, the volume is about 36%.

Extrusion is a process that should run continuously. Industrial extruders are run 24 hours a day during a work week. They are shut down and cleaned over the weekend.

Industrial Extrusion Equipment

Industrial extruders are many sizes. The barrel diameter ranges from 1-1/2″ to 15″ (38 mm to 380 mm). Some experimental machines have been made. Some had barrel diameters as large as 36″ (915 mm). Others had two, three, four, and even five screws built into one machine.

Sterling Extruder Corp.

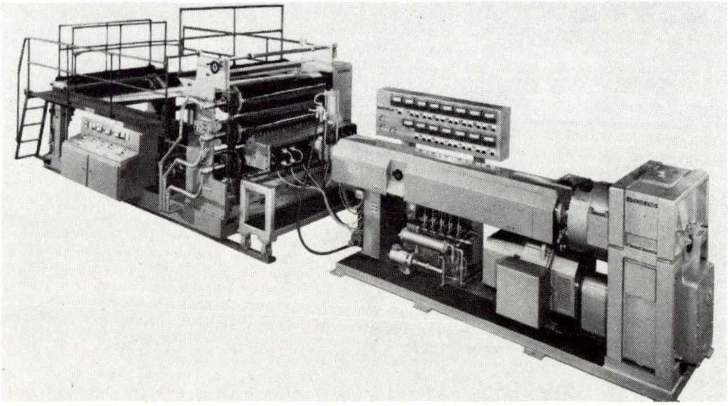

Fig. 2-20. Sheet extrusion line set up for use.

Materials Used in the Industry

Most thermoplastic materials may be extruded. Among the most important are polyethylene, polystyrene, polypropylene, polyvinyl chloride (PVC), and the cellulosics.

As thermoplastic materials are melted in the extruder, they pass through a physical change. When they go through the cooling device, another physical change occurs. As they are cooled, they solidify.

BLOW MOLDING

Process Description

Blow molding is a process for forming hollow objects from thermoplastics. Products such as bottles, garbage cans, and toys are common.

Three methods of blow molding are used today. They are the direct, the indirect, and the injection methods.

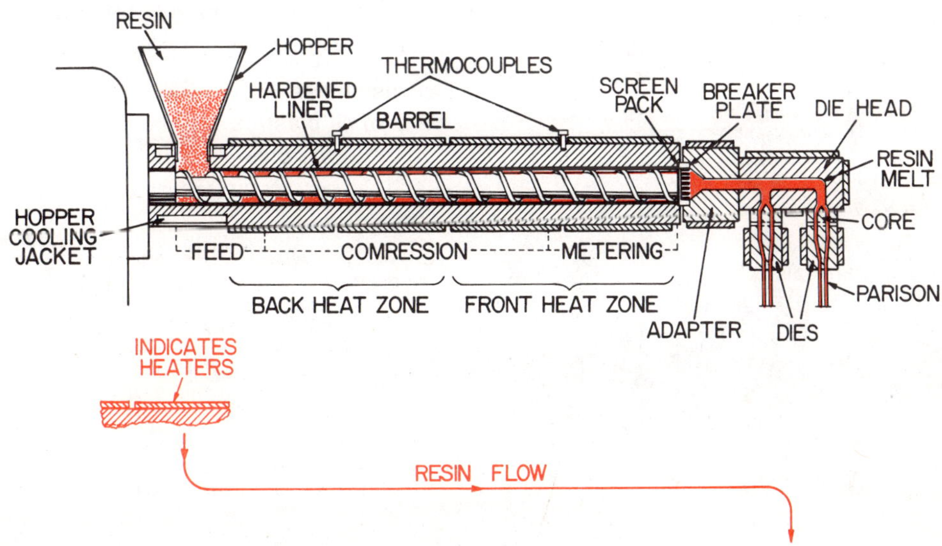

Fig. 2-21. Cross section of an extrusion blow molder with twin die heads.

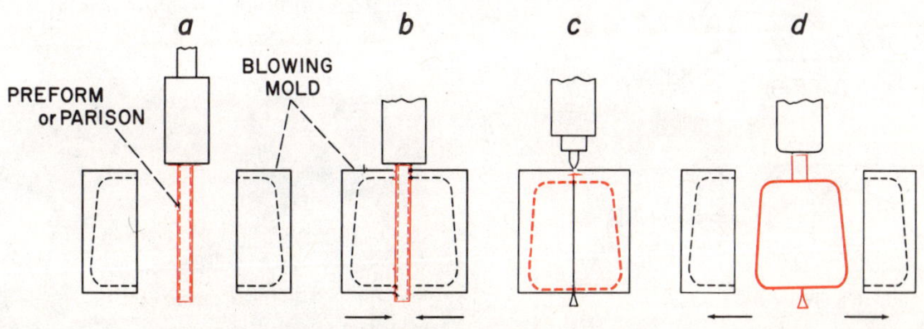

U.S. Industrial Chemicals

Fig. 2-22. Blow molding — direct method.

A. Parison extruded.
B. Molds close.
C. Parison blown by air to shape of mold.
D. Mold opened and part removed.

Direct Method. The direct method of blow molding is often known as **extrusion blow molding.** It probably produces the largest amount of blow molded products. In this method, a hollow plastic parison (roughly shaped tube) is extruded between two open mold halves. The plastic parison extends to the bottom of the mold. The mold closes. Air is blown into the center of the parison, and the plastic takes the shape of the mold.

High-speed extrusion blow molding equipment produces hollow containers with open necks. Some of these are detergent bottles, milk bottles, gas cans, and tanks.

Indirect Method. Products may be formed by the indirect method of blow molding in two ways. These are: (1) the tubing method and (2) the two-sheet method.

In the **tubing method,** plastic tubing is extruded and cut to length before use. An air hose is placed in one end. Then the tubing is heated and clamped between the mold halves. Air is blown through the hose into the plastic. The plastic takes the shape of the mold. It is cooled, removed, and trimmed. Long, hollow objects, such as hosiery display legs, are often blown by the indirect tubing method. See Fig. 2-24.

In the **two-sheet indirect method,** heat and pressure are applied at the joint. This welds the

Rainville Co.

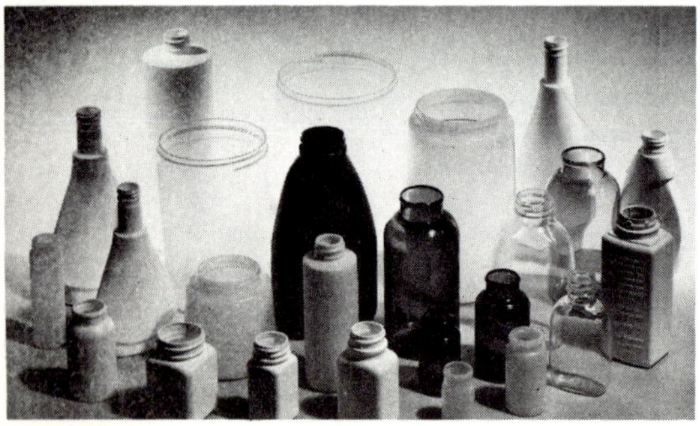

Fig. 2-23. Blow molded products.

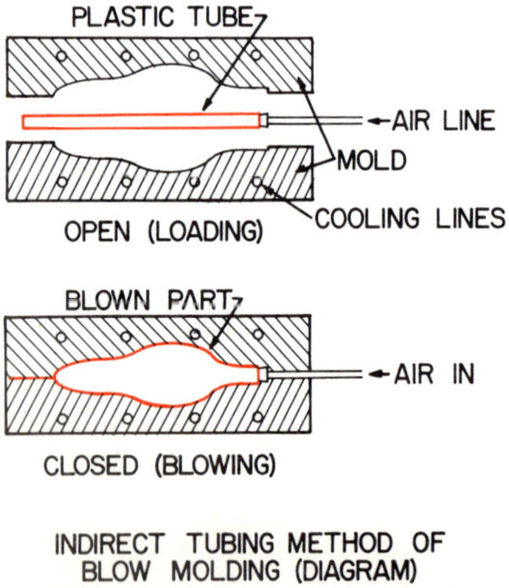

Fig. 2-24. Indirect tubing method of blow molding.

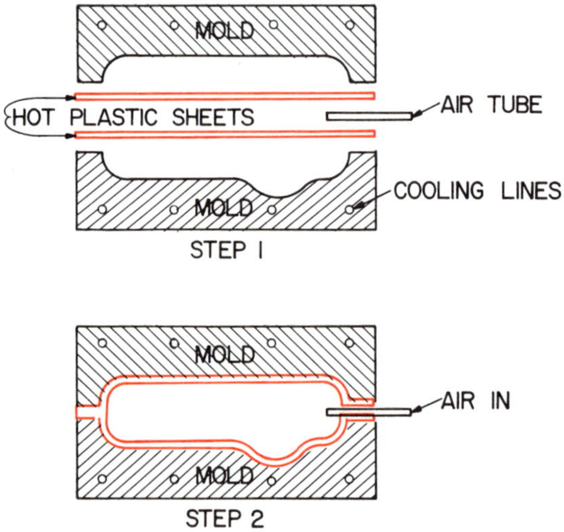

Fig. 2-25. Indirect two-sheet method of blow molding.

Fig. 2-26. Toy boat formed by indirect two-sheet method of blow molding.

two plastic sheets together. Large hollow signs or tanks may be made by this method. See Figs. 2-25 and 2-26.

Injection Blow Molding. During injection blow molding, a parison, or hollow plastic slug, is first injected over a mandrel (supporting bar). The mold halves are opened. The parison is moved to the blow mold halves. The halves close over the parison which is still on the mandrel. The object is blown. See Fig. 2-27.

Injection blow molding is the newest form of blow molding. It will make a completely finished part without flash (plastic escaping at parting lines). No trimming is required as in extrusion blow molding. Parts made by this method have very even wall thicknesses. Injection blow molding does not allow production of products with handles or irregular shapes.

Importance, Advantages, and Disadvantages of Blow Molding

Blow molding now processes about 4% of all thermoplastic materials. A rise to 6% or more is expected by 1980. More products will be packaged in plastics. Blow molded plastics are rapidly replacing glass, metal, and paper containers.

Blow molding provides a quick method of processing bottle-like containers with open ends. Mold costs are lower than those for injection molding and compression molding. Blow molded parts — from very small to quite large — are possible. The scrap plastic may be recycled. Blow molding can usually be adapted to common extrusion equipment. Newer blow molders, however, are built around specially designed extruders.

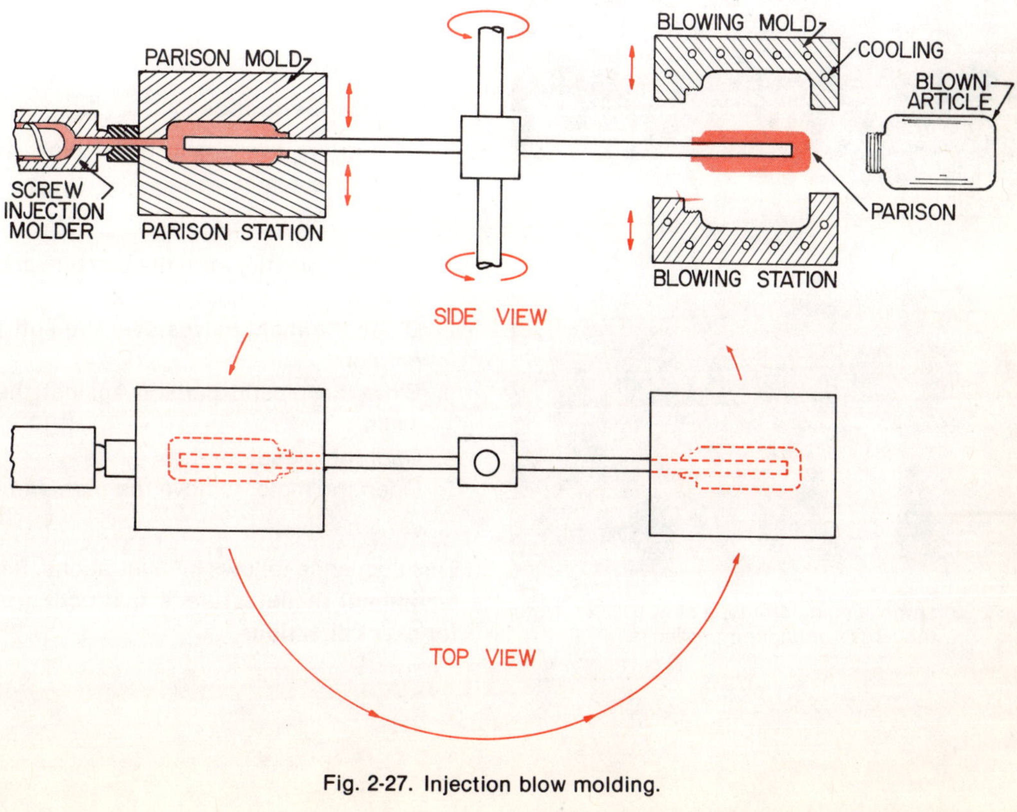

Fig. 2-27. Injection blow molding.

Rainville Co.

Fig. 2-28. Injection blow molding machine.

Plastics Machinery & Equipment Magazine

Fig. 2-29. Large accumulator-type blow molder. Note the large container it produces.

Industrial Equipment

Blow molding machines are sold by the size of the blow molded product they produce. Product container sizes range from one ounce (29 ml) to 66 gallons (250 liters). The blow molder size is sometimes given by its extruder or injection screw size. It ranges from 1-1/2″ (38 mm) to 6″ (152 mm) in diameter.

Materials Used in Industry

Almost any thermoplastic can be blow molded. The most common is polyethylene. High density polyethylene is the more popular. It is used for the more rigid containers. Milk and bleach bottles are examples. Cellulosics, polyvinyl chloride, polycarbonate, acrylic, ionomer, and polyacetal plastics are also used for blow molding. Polypropylene is becoming a popular blow molding resin. It makes clearer bottles than polyethylene.

Operation Sequence

There are five steps in the blow molding process. These are:

1. Heat the plastic, form the parison, and place it between the mold halves.
2. Close the mold halves over the soft plastic parison.
3. Blow the plastic parison against the mold walls.
4. Cool the plastic part.
5. Open the mold, remove the part, and trim it.

This sequence follows in illustrations. **Read the equipment manufacturer's instruction manual for exact directions.**

1A

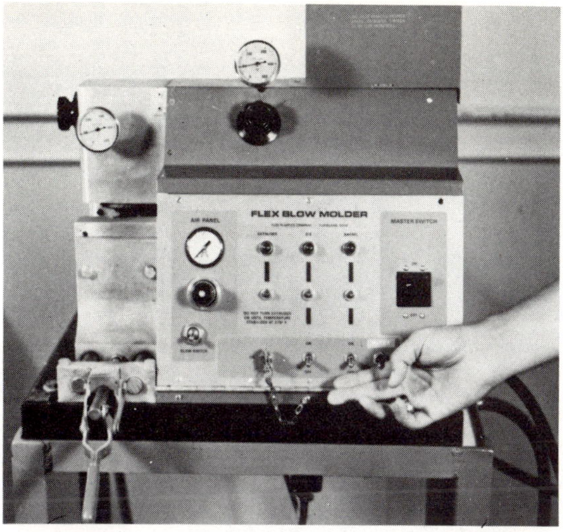

Fig. 2-30. Turn on heating units.

1A Turn on the barrel and die heaters. Bring them up to operating temperature. From 50° F to 100° F (27° C to 55° C) above the plastic's melting point is needed. **Caution: Do not turn on extruder screw or plunger before barrel and die reach operating temperatures.**

1B

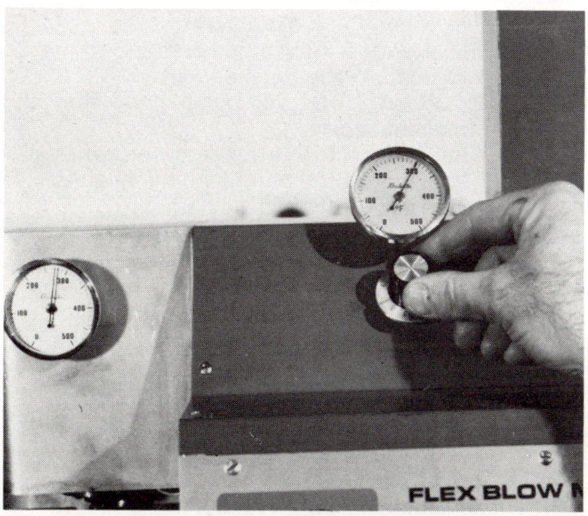

Fig. 2-31. Adjust barrel and die heaters.

1B **Do Not Turn Heater Regulator Knobs to Highest Settings.** Select a medium setting and adjust temperature after heater indicator lights go out. Check heater thermometers. **Read the manufacturer's directions for exact preheating sequence.**

1C

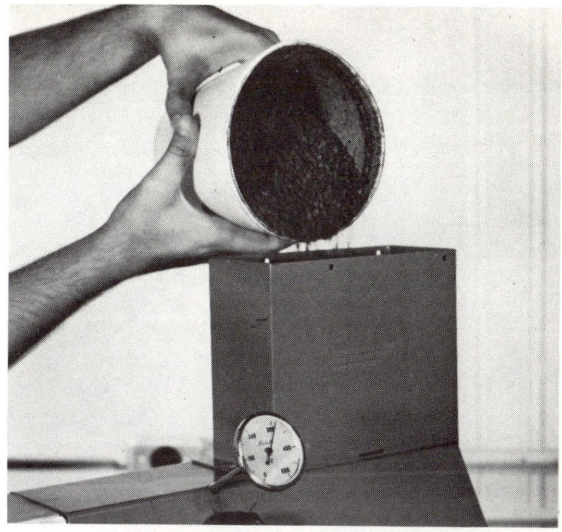

Fig. 2-32. Fill hopper.

1C Fill the hopper with material. Use a low melt index (1.5 to 2.5) polyethylene to start with. **Read machine manufacturer's directions for other materials.**

Adjust the parison blowing pressure regulator to about 10 psi (69 kPa).

2A

Fig. 2-33. Extrude the parison.

2A (a) Be sure barrel and die head are at operating temperatures. Extrude several lengths of molten plastic with both mold halves open. Check for uniform parison.

(b) Remove trial parisons with a wood stick. Do not touch the hot plastic. It will burn you.

(c) Extrude a parison 1/4″ to 1/2″ (6 mm to 13 mm) below the mold pinch-off.

2B

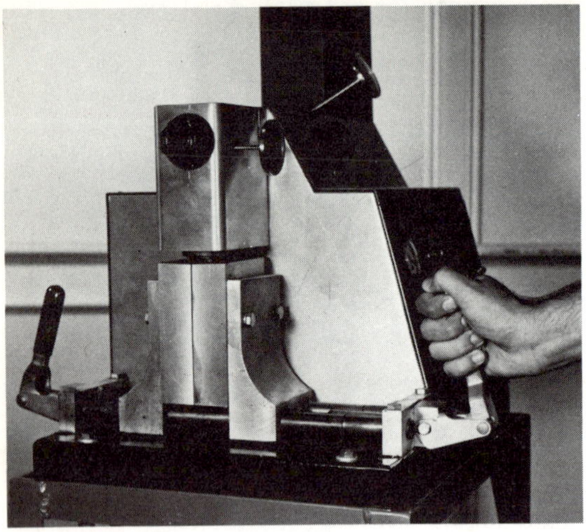

Fig. 2-34. Clamp the mold shut.

2B Close both mold halves fully.

3

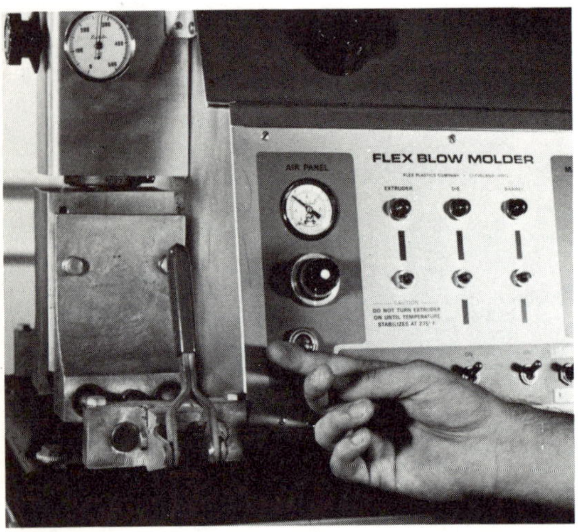

Fig 2-35. Turn on the blowing air and cool the part.

3 Turn on the blowing air for 10 to 15 seconds.

4 Allow the part to cool in the mold for 20 to 30 seconds. As the mold gets warm, allow a longer cooling time.

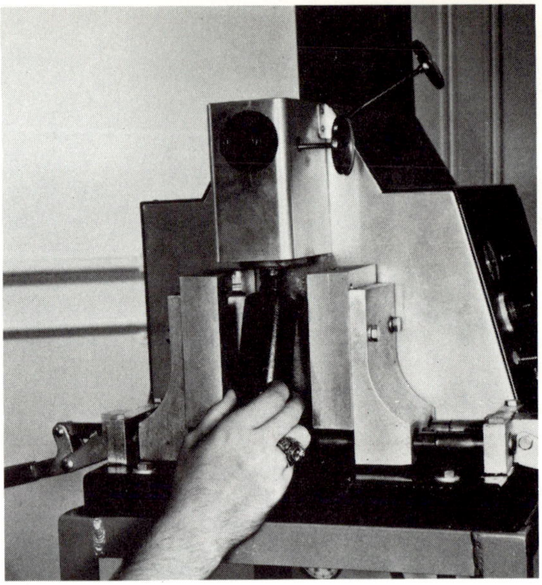 5

Fig. 2-36. Remove part.

5 (a) **Put on a glove to protect your hand during this step.**
(b) Open the mold halves.
(c) Remove the molded part and trim it.
(d) Inspect the finished part.

When parts are completed, shut the machine down.
(a) Turn off the extruder motor switch.
(b) Turn off the barrel and die heaters.
(c) Turn off the blowing air.

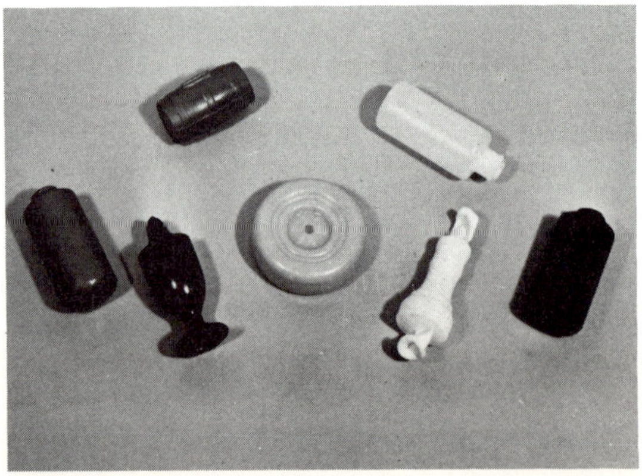

Fig. 2-37. Products made on laboratory blow molders.

STUDENT ACTIVITIES

1. Process a variety of plastics in each available mold.
2. Experiment with plastics having different density and melt index. **Caution: Stay within the blow molding melt index range.**
   a. Try several different melt index resins for a given mold.
   b. Mix two **blow molding grade** polyethylene resins together. Blow mold them. Note how the blend averages to a density between the two parent materials.
3. Experiment with the blow molding process. Change **only one** variable at a time. Start with normal established settings for a given material and machine. Do the following:
   a. **Increase** the blowing time one second with each part.
   b. **Decrease** the blowing time one second with each part.
   c. Blow several parts, **increasing** the blowing pressure 1 psi (6.9 kPa) for each part.
   d. Blow several parts **decreasing** the blowing pressure 1 psi (6.9 kPa) for each part.
   e. Increase the barrel and die temperatures slightly (10° F or 5° C). Mold several parts. Note what changes occur in the parts. Increase another 10° F (5° C) and repeat the above. Do not go more than 50° F (25° C) above the normal settings in this experiment.
4. Mix dry color pigments with the plastic pellets. Produce a variety of colored products. Various colors may be mixed with each other also.

COMPRESSION AND TRANSFER MOLDING

Process Description

Compression and transfer molding processes compress plastic materials into shape. The materials are cured or hardened in a metal mold. Both thermoplastic and thermosetting plastics may be molded in this way. Thermosetting plastics are usually used.

A variety of products are made by the compression molding process. These include electric toaster handles, light-switch cover plates, and plastic dishes. Common transfer molded parts include pot and pan handles, auto distributor caps, and electric switch parts.

Compression Molding

Compression molding machines operate in the following manner. Plastic materials are measured by weight or volume. They are placed in the open **mold cavity** (1). See Fig. 2-38. The mold is usually heated from 290° to 400° F

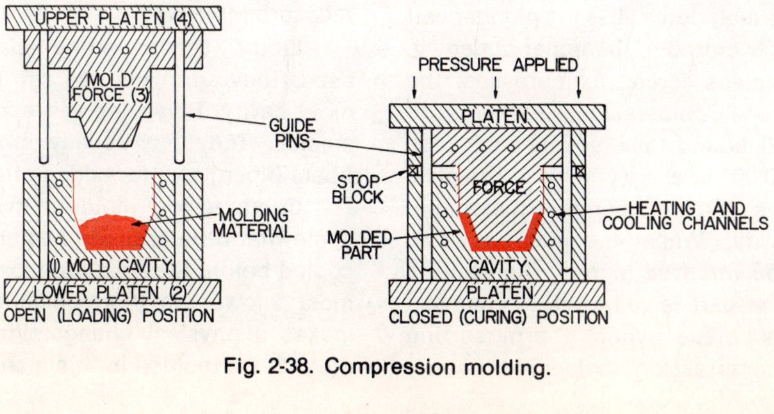

Fig. 2-38. Compression molding.

Durez Div., Hooker Chemical Co.

Fig. 2-39. Pellets and granular thermosetting plastics.

Fig. 2-40. Compression and transfer molding preforms.

(143° C to 205° C). The mold cavity is usually bolted to the **lower platen** (2) of the compression press. The **mold force** (3) is the plunger half of the mold. It is bolted to the **upper platen** (4). One of the platens forces (compresses) the halves together. Compression pressures of 1000 to 12,000 pounds per square inch (psi) (7000 kPa to 80,000 kPa) are common. The plastic melts, flows, and then cures into shape in the mold cavity. When the thermosetting material has **polymerized** (cured), the mold is opened and the part is removed. A **chemical change** takes place when thermosetting plastics are compression molded. It usually

Plastics Engineering Company

Fig. 2-41. Compression mold used for thermoset molding compounds.

takes from 1 to 20 minutes for this chemical change to occur.

The plastic materials placed in the mold cavity may be in granular or pellet form. See Fig. 2-39. Or they may be formed into tablets called preforms. Examples of these are shown in Fig. 2-40. Phenolic, melamine, urea, and DAP (diallyl phthalate) plastics are often premeasured this way.

Liquid thermosetting plastics may be used. They are measured and poured into the mold cavity. Polyester and epoxy are two examples. They are usually mixed with glass fibers (fiberglass) for added strength.

Compression molding thermoplastics is somewhat different. The heated mold must be cooled before the part is removed. Cooling the mold allows the thermoplastic to harden. It makes a **physical change.** Vinyl phonograph records are molded in this manner.

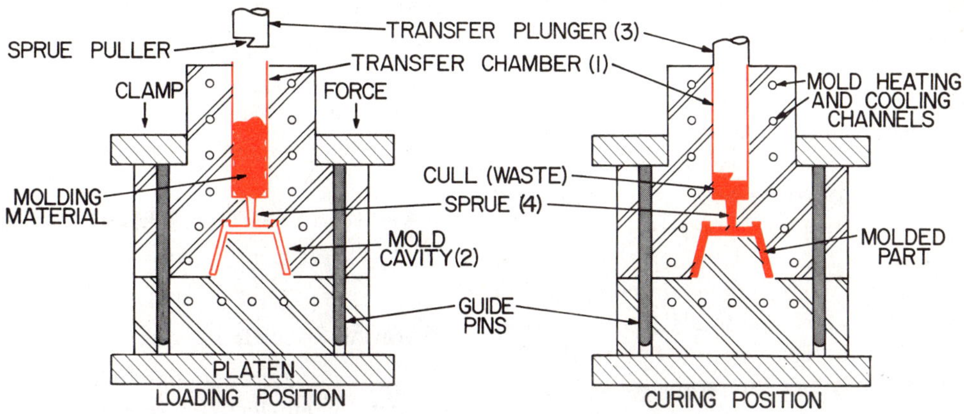

Fig. 2-42. Transfer molding — sprue puller design.

Transfer Molding

Transfer molding is a thermosetting molding process. It is similar to plunger injection molding. Transfer molding is associated with compression molding. It uses similar equipment and materials. Products made are similar to those typically compression molded.

A diagram of the transfer molding process is shown in Fig. 2-42. Thermosetting plastic is placed into a heated **transfer chamber** (1). Here, it is heated. The plastic is then transferred (forced) into a heated, closed **mold** (2). This is done by the **transfer plunger** (3). Transfer molding pressure (transfer plunger pressure) is usually twice that of compression molding. A pressure drop between the **sprue** (4) and the mold cavity makes this necessary. Pressure in the mold is similar to compression molding. Mold temperatures of 320° to 350° F (160° to 175° C) are common. Polymerization (cross-linking, a chemical reaction) takes place in the mold cavity. After polymerization, the mold is opened and the cured part is removed.

Thermosetting materials are weighed or measured for transfer molding. They may be in a preform, pellet, or granular form.

Importance, Advantages, and Disadvantages of Compression and Transfer Molding

Compression and transfer molding combined are about fifth in importance in the plastics industry. About 4% to 5% of the plastics materials are processed by these methods yearly.

Plastics Engineering Company

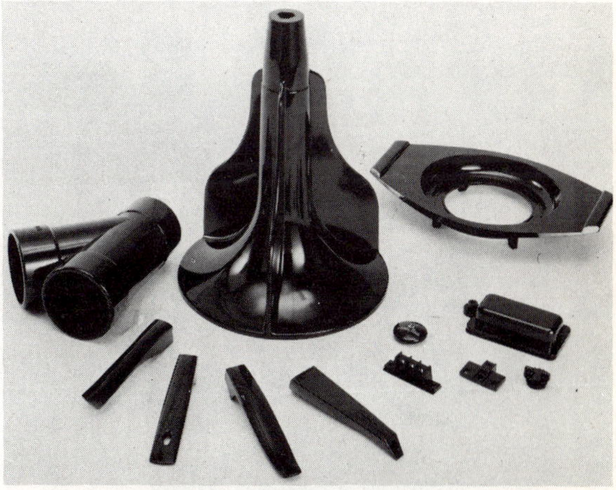

Fig. 2-43. An assortment of parts molded from black phenolic thermoset molding compounds. They have heat- and impact-resistant properties.

Many parts will probably be compression or transfer molded for years. Plastic housewares often require hard, scratch-resistant surfaces. Compression molded thermosetting plastics have these qualities. Plastic dishes, household electrical switches, electrical outlets, and light fixture parts are made in this manner.

Fiber-reinforced thermosetting plastics develop maximum strength when compression molded. Transfer molding and injection molding tend to break up the fibers. This reduces the strength of the product.

Transfer molding offers many of the advantages of compression molding. Also, holes and inserts can be molded in easily. Transfer molding is often used for parts such as distributor caps, ignition rotors, and switch parts. Multiple-cavity molds may be economically used in transfer molding. Thermosetting materials for transfer molding and compression molding cost less than comparable thermoplastics.

Durez Div., Hooker Chemical Company

Fig. 2-44. Typical transfer molded parts.

Materials Used in Industry

Many thermosetting plastics may be compression and transfer molded. Phenolic, melamine, urea, alkyd (polyester), diallyl phthalate (DAP), and epoxy are commonly used.

LABORATORY COMPRESSION MOLDING

Material Specifications

General purpose phenolic thermosetting material is usually used for laboratory compression molding. A medium flow (8-12) wood flour-filled phenolic works well. It should have low to medium bulk factor (2.0-3.5).

Operating Pressures

Operating pressures vary. They depend on the material and mold used. A general rule to follow for pressures is given in Fig. 2-45.

Operation Sequence

There are six steps in the compression molding process. These are:

1. Preheat the mold to the right temperature. Coat the mold with mold release.
2. Weigh out the amount of plastic needed.

Material	PSI (kg/cm²) for Surface Area[1]	P/I (kg/cm) for Sides[2]
Phenolic	2000 (140 kg/cm²)	1000 (178 kg/cm)
Urea	3000 (210 kg/cm²)	1500 (268 kg/cm)
Melamine	4000 (280 kg/cm²)	2000 (357 kg/cm)

[1] Pounds (kilograms) of force for each square inch (kg/6.45 cm²) of flat molding surface area.

[2] Additional pounds of force for each inch (kg/cm) of side height (pounds per inch of side height or kilograms per centimeter side height).

Fig. 2-45. Operating pressure for compression molding.

3. Fill and close the mold. Put it in the press.
4. Apply enough force (pressure) to the mold to make a compact part with good surface finish.
5. Release compression pressure to allow gases to escape. Reapply molding force.
6. Let the part cure in the mold.
7. Open the mold and remove the part.

The following illustrations show the operation sequence. Follow it carefully. **Read the manufacturer's instruction manual for exact specifications and instructions.**

1A

Fig. 2-46. Heat the platens and mold.

1A Plug in the electrical cord. Turn on both platens, and allow the mold and platens to heat up together. 300° F to 350° F (149° C to 177° C) is normal.

1B

Fig. 2-47. Wax the mold.

1B (a) Remove the mold from the platen. **Caution: Hot. Use gloves.**
(b) Wax the heated mold with carnauba wax.
(c) Be sure to have a heat-resistant (transite) cover on wood bench tops for compression molding or use steel-top benches.

2

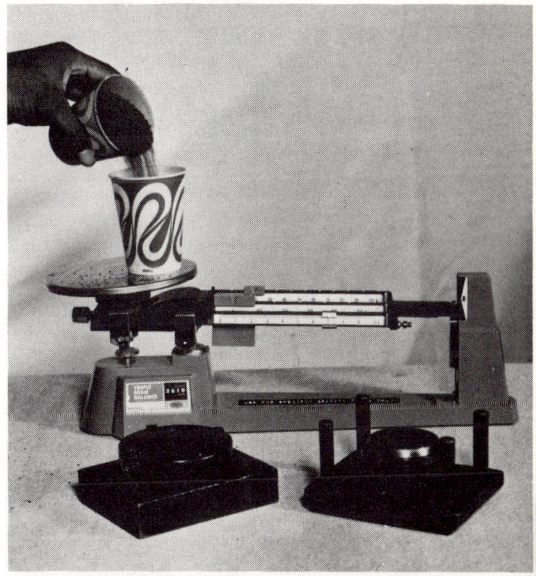

Fig. 2-48. Weigh the material charge. (Mold is in the foreground.)

2 Weigh out the proper amount of material for the mold. Experience with a mold will determine the exact amount of charge.

3A

Fig. 2-49. Fill the mold.

3A (a) Pour the charge of material into the mold.
(b) Distribute it evenly in the mold.

3B

Fig. 2-50. Close the mold.

3B It should close only one way. Match up the index marks for proper closing.

3C

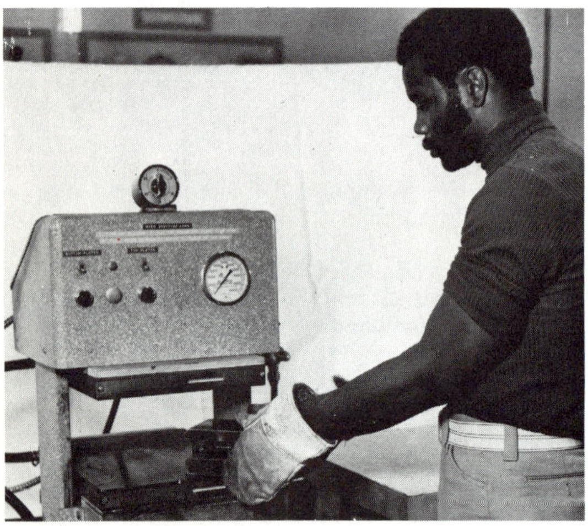

Fig. 2-51. Place the mold on the lower platen.

3C Place it in the center, both front to back and side to side.

 4

Fig. 2-52. Compress the material.

4 Apply pressure as figured according to Fig. 2-45.
Note: Pressure will not build up rapidly at first.

5 "Breathe" the mold cavity by releasing the molding pressure. Build the molding pressure back up again until the part is cured.

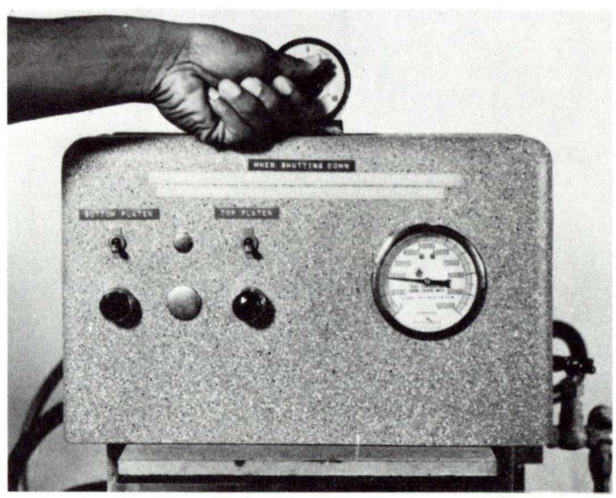 6

Fig. 2-53. Set the timer.

6 Hold this pressure for several minutes. Start with 5 minutes at 350° F (177° C). Some parts may be cured in a shorter time.

7
(a) Open the press and remove the mold.
(b) Remove the part from the mold.
(c) Allow the part to cool.
(d) Inspect the part for defects.
(e) Trim the flash from the edges with a file or sandpaper.

Safety Precautions

Hot platens and hot molds **must** be guarded. Do this so accidental burns will not happen. **Safety zones** around the machines and **hot signs** are necessary. When machines are shut down, hang hot signs in front of the platens. See Fig. 2-54.

Hot molds should be waxed with **carnauba** wax. Place them inside of a bench or cabinet or back out of the way. **Do not place on a wooden shelf as a fire may result.** Bench tops and shelves around compression molders should be metal or covered with asbestos-cement (Transite) board.

Molds should be left open when cooling. This keeps them from rusting.

Fig. 2-54. Hot sign on the platens.

STUDENT ACTIVITIES

1. Process a variety of thermosetting materials with prepared molds.
   a. Phenolics.
   b. Melamine.
   c. Urea.
   d. Genal® or Durez Si.®
   e. Polyester/fiberglass.
   f. Powdered pure phenolic and wood chips.
2. Experiment with materials in prepared molds. Make the following variations:
   a. Material types.
   b. Colors.
   c. Mix colors of the same material type.
   d. Change bulk factors.
   e. Mix in filler such as powdered wood (sanding dust) or ground glass fibers (1/8″ or 3 mm long). Compare with products made of unmodified resin.
   f. Change flow numbers.
   g. Mix two materials of the same type but with different flow numbers. The flow characteristics should be somewhat between the two parent materials.

FORMING AND LAMINATING

Chapter 3

Forming and laminating includes thermoforming, laminating, and roll forming. All are groups of processes.

Thermoforming processes use hot plastic sheet or film stock. The sheet or film is formed over or into a mold by vacuum, air pressure, or mechanical force. There are two kinds of vacuum forming. These are drape forming and cavity forming.

Laminating processes use layers of plastic materials or plastics combined with other materials to form products. It includes high- and low-pressure lamination. Low-pressure laminated products are made from fiberglass or thermoplastic sheets. High-pressure lamination can be divided into two groups: (1) high-pressure fiberglass lamination and (2) high-pressure thermosetting lamination.

Roll forming is a group of processes in which the plastic is formed using rollers. The plastic is either formed onto or into a sheet. The sheets may be backed with a **substrate** (fabric, paper, or fiberglass material). Or they may be unbacked. Roll forming includes calendar

NRM Corp.

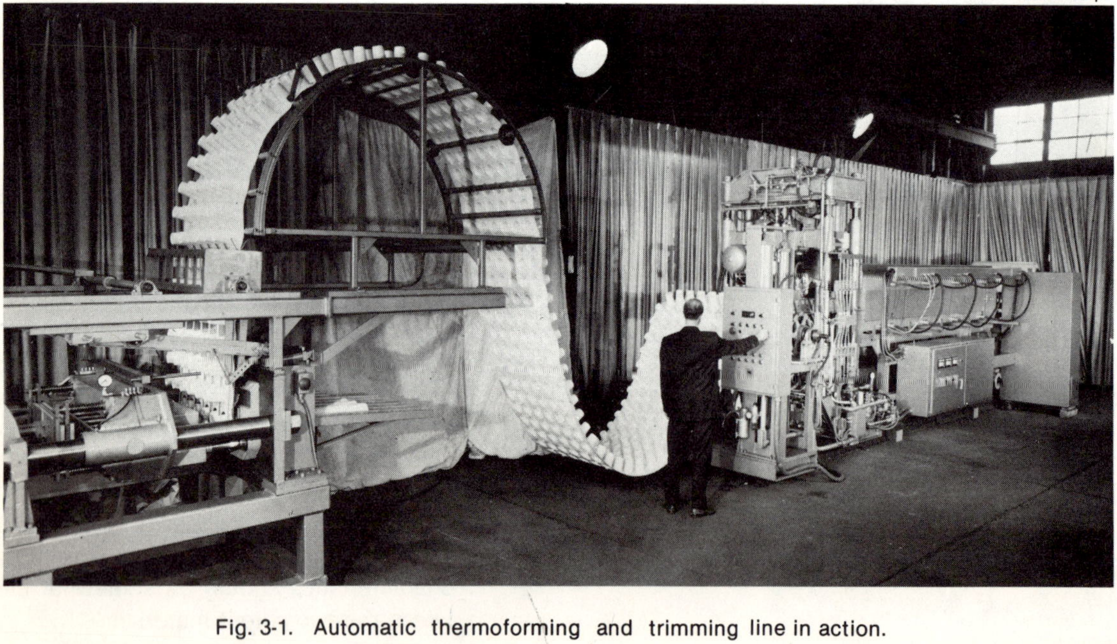

Fig. 3-1. Automatic thermoforming and trimming line in action.

coating, knife (spread) coating, and extrusion coating.

Forming and laminating may be compared with molding. Parts made by forming and laminating are often larger. Some forming and laminating production rates are lower than molding rates. Equipment and molds for forming and lamination are often less costly.

THERMOFORMING

Process Description

Thermoforming plastics processes use heated plastic film or sheet. The film or sheet is formed over or into a mold. The plastic sheet hardens as it cools on the mold. It will hold the shape of the mold. The hardening is a **physical change.**

Thermoforming is divided into four categories.

1. Vacuum forming (two types).
   a. Drape vacuum forming.
   b. Cavity vacuum forming.
2. Pressure forming.
3. Free blowing.
4. Mechanical stretch forming.

Vacuum Forming

A vacuum is created when atmospheric (air) pressure is removed. In vacuum forming, a heated plastic sheet is sealed against the mold. A vacuum is drawn (pumped) on one side of it. Atmospheric pressure on the other side forces the plastic against the mold. See Fig. 3-2. Vacuum forming does not work as well at high altitudes. The atmospheric pressure is lower.

Vacuum forming is probably the most used thermoforming method. It is divided into two kinds. One kind is **drape vacuum forming.** A heated thermoplastic sheet is formed over a male mold. The plastic cools and hardens on the mold. The formed plastic part is then taken off the mold. Extra plastic is trimmed off.

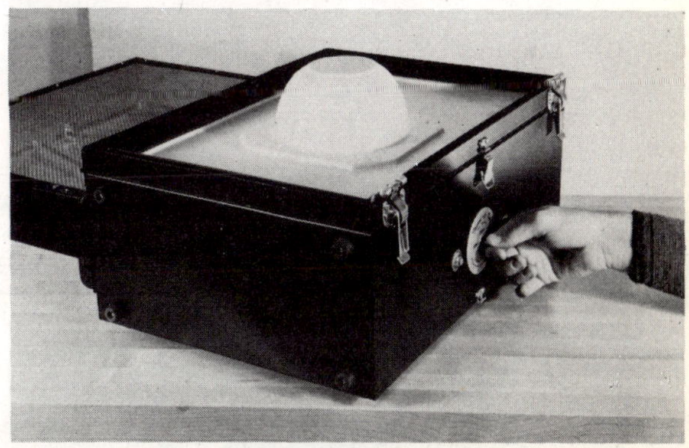

Fig. 3-3. Drape vacuum forming.

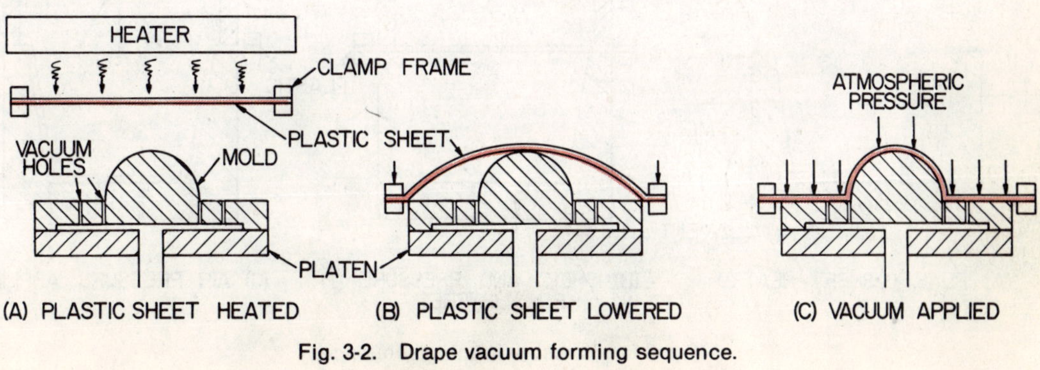

Fig. 3-2. Drape vacuum forming sequence.

The other kind of vacuum forming is **cavity vacuum forming.** A heated thermoplastic sheet or film is formed into a female or cavity mold. It is cooled and taken from the mold as in drape forming. It is also trimmed before use.

Fig. 3-4. Cavity vacuum forming.

In most thermoforming, a clamp frame is used. It holds the thermoplastic sheet tightly on all four sides. This keeps the plastic from shrinking when heated. It also helps the plastic form a tight vacuum seal on the mold.

Pressure Forming

In pressure forming, air pressure is used. The thermoplastic sheet is "blown," onto or into the mold. First, the thermoplastic sheet is heated. A pressure box or mold is usually lowered over it. Air pressure is added to the box. This forces the plastic sheet against the mold surface. See Fig. 3-6.

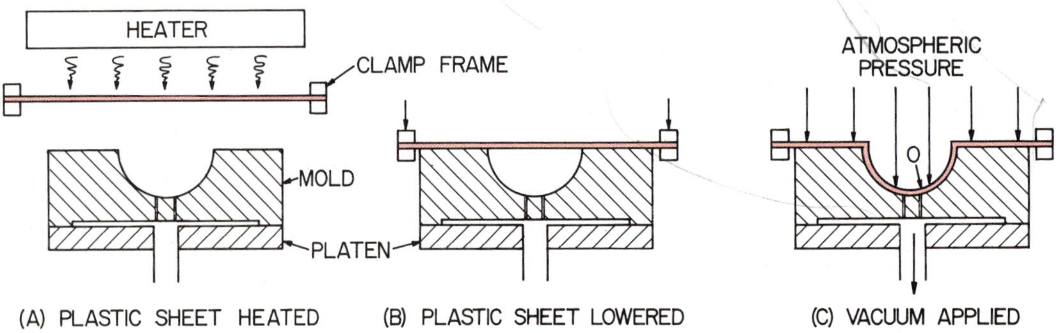

Fig. 3-5. Cavity vacuum forming sequence.

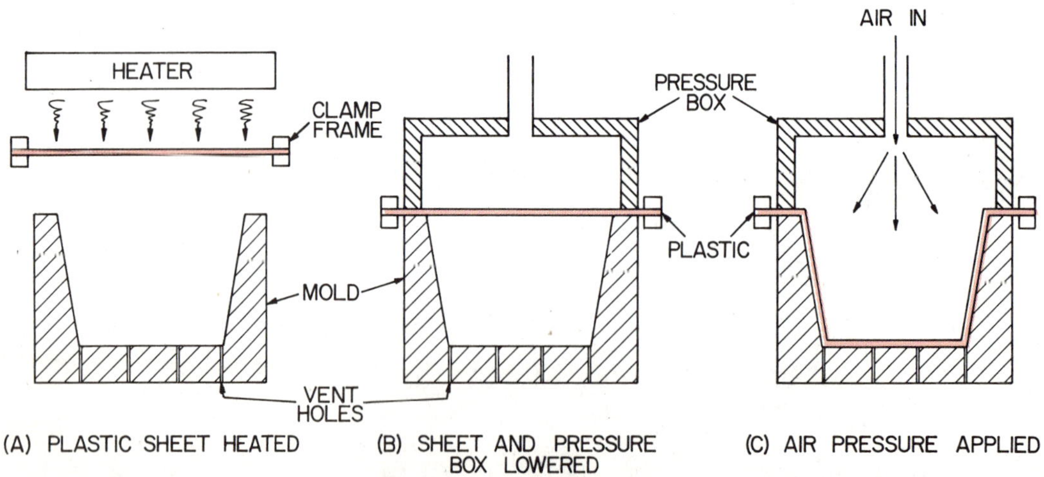

Fig. 3-6. Pressure forming sequence.

Pressure forming is often used for thick or stiff thermoplastic sheet. It also speeds up production of thin thermoformed parts. More pressure is used than in the vacuum forming method.

Free Blowing

Free blowing is another method of thermoforming. It is used to form parts which can be seen through. No mold is used. The hot plastic sheet is clamped between a ring and platen. Air is blown through a hole in the platen. The plastic blows up like a dome. Acrylic astrodome panes, serving tray covers, and other such parts are formed this way.

Simple equipment for free blow forming may be built in the school laboratory. It is used to form acrylic plastic sheet. See Fig. 3-7. Free blowing may also be done on laboratory and commercial thermoforming equipment.

Fig. 3-7. Simple shop-made free blowing equipment.

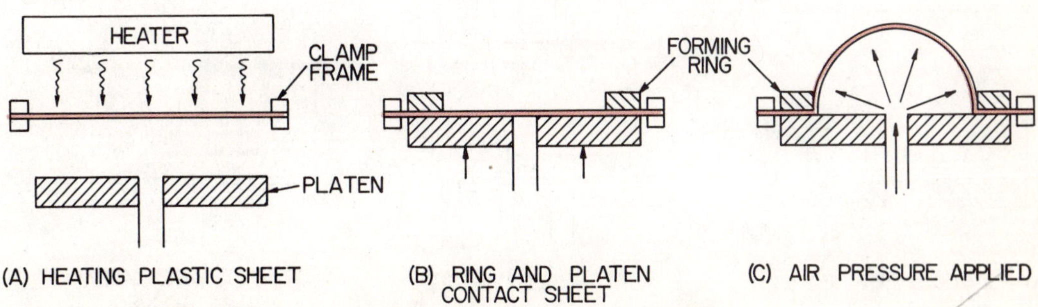

Fig. 3-8. Free blowing sequence.

Mechanical Stretch Forming

Mechanical stretch forming is also called plug forming. This thermoforming method uses mechanical pressure to form the sheet. Usually, air or hydraulic cylinders move the mold(s) to stretch or form the thermoplastic sheet. See Fig. 3-9. The mold(s) could also be gear, lever, or screw operated.

Mechanical stretch forming serves other purposes as well. It is used to form thick or stiff plastic sheet. It may also help in vacuum or pressure forming. The plug stretches the thermoplastic sheet. Then the vacuum or air pressure is applied. Mechanical stretch forming helps make parts with even thickness.

Process Combinations

Combinations of two or three thermoforming processes may be used. Pressure forming and vacuum forming are often combined. Mechanical stretch forming may be added to either of these processes or their combination. The plastic may also be stretched by compressed air before being formed. A plug

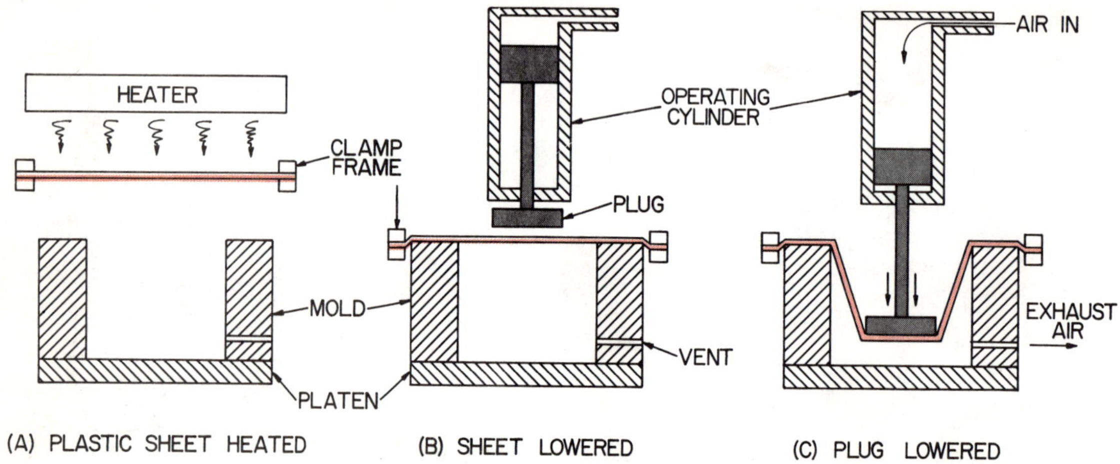

Fig. 3-9. Mechanical stretch forming sequence.

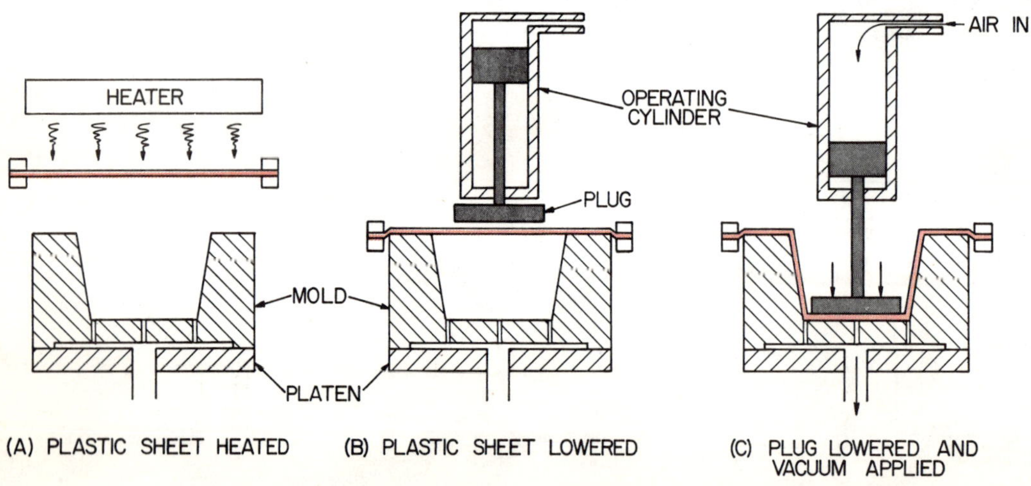

Fig. 3-10. Plug assist thermoforming sequence. A process combination.

(mechanical stretch forming) may push it into the mold. It can then be vacuum formed. There are a number of combinations of thermoforming. New ones are still being developed. Remember, any two or all of the thermoforming processes may be combined.

Thermoforming Techniques for Packaging

Many products are packaged in see-through plastic. It is inexpensive. It also permits easy consumer selection. Some methods of packaging use the thermoforming technique.

Blister Forming. Blister forming is a vacuum forming technique. It is used to make package covers for products. Clear plastic sheet or film may be formed by drape or cavity vacuum methods. A shape slightly larger than the product to be enclosed is made. The plastic blisters or domes are cut from the sheet. They are then attached to printed card stock with the product enclosed.

Skin Packaging. In skin packaging, a product is placed on a special card in the thermoformer. See Fig. 3-12. A thin ionomer or polyethylene film is vacuum formed over the product. The card stock is perforated. This allows air to pass through it. The plastic film sticks to the card stock and holds the product in place.

Materials Used in Industry

Any thermoplastic material that can be made into sheet or film may be thermoformed. High-impact polystyrene sheet is a popular thermoforming plastic. Polyethylene film and sheet are thermoformed into package covers. Clear package covers are made from cellulosics, such as cellulose acetate. Polyvinyl chloride (PVC or vinyl) film and sheet are thermoformed into clear and colored packages. Thick and/or strong products are thermoformed from ABS.

Importance, Advantages, and Disadvantages of Thermoforming

Thermoforming is an important plastics process. Many products can be made inexpensively. Large advertising signs, such as lighted gasoline station signs, are thermoformed. Many packages are made this way. Future automobile bodies may also be thermoformed. Machine and tooling costs are low. Thermoforming is likely to be one of the most important plastics processes of the future.

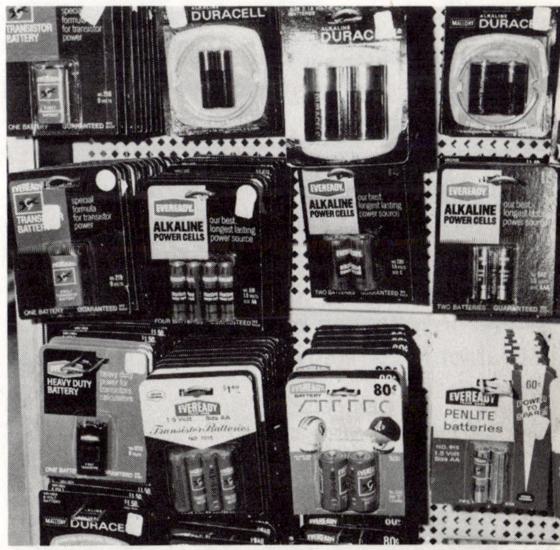

Fig. 3-11. Display of blister packaged batteries.

American Packaging Corp.

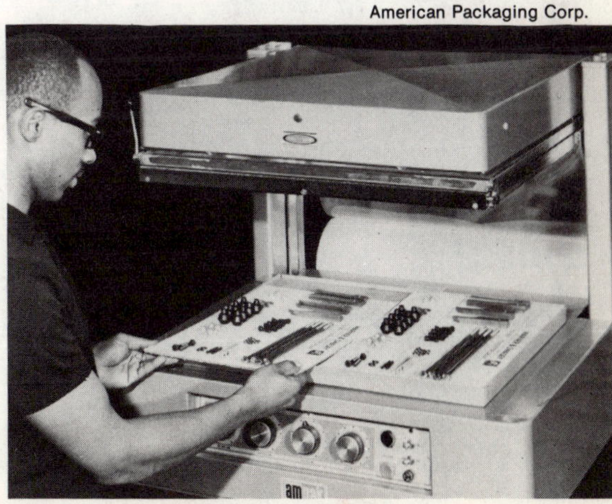

Fig. 3-12. Skin packaging.

Laboratory Thermoforming

Plastics most often used for laboratory thermoforming in schools are high-impact polystyrene, cellulose acetate, ABS, polyethylene, and ionomer.

Operation Sequence

The thermoforming process has nine steps.

1. Install the mold on the platen.
2. Cut the plastic sheet to size.
3. Clamp the plastic sheet or film into the thermoformer.
4. Heat the plastic until it begins to sag.
5. Bring the plastic into contact with the mold.
6. Apply the vacuum or pressure, or both, as needed.
7. Cool the formed plastic part.
8. Remove the part from the press.
9. Trim the part as needed.

The following illustrations show the operation sequence in thermoforming. **Read the machine manufacturer's instruction manual for exact directions.**

(See Safety Notes on page 42.)

1

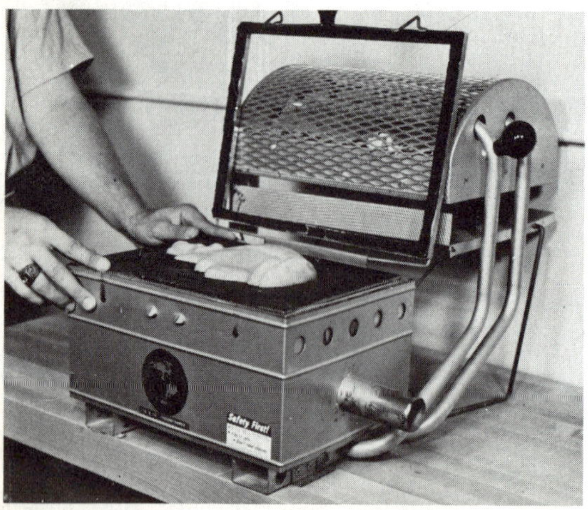

Fig. 3-13. Mount mold in press.

1 Put the mold(s) on the platen(s) of the thermoforming machine. Adjust the mold(s) so they center under the clamp frame. Fasten the mold(s) if necessary.

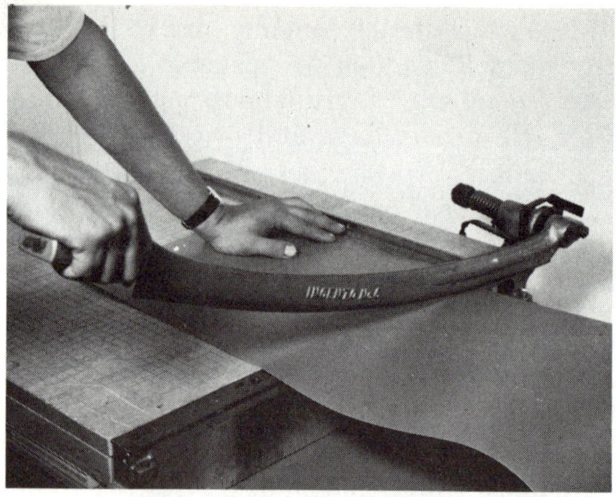

2

Fig. 3-14. Cut plastic to fit frame. Thicker sheet can be cut on a sheet metal shear or table saw.

2 Cut the plastic sheet to the right size for the clamp frame. Sheet should be slightly larger than the clamp frame in most cases. This operation is not needed for roll feed machines.

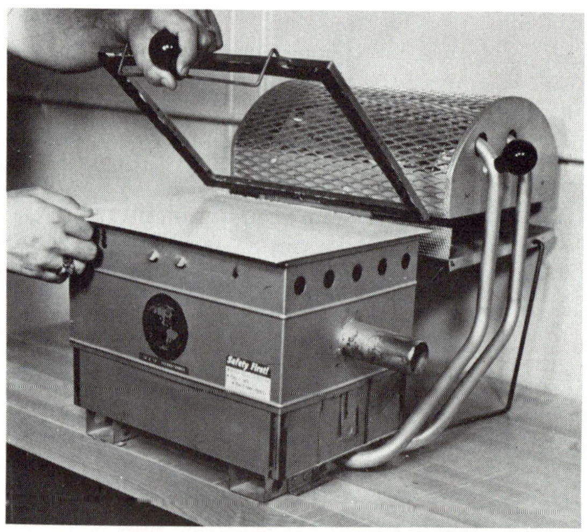

3

Fig. 3-15. Put plastic in clamp frame.

3 Center it well. Narrow frames usually need at least 1/4″ to 1/2″ (6 mm to 13 mm) extra plastic around the frame. **Example:** For a 10″ x 14″ (254 mm x 356 mm) thermoformer, cut the sheet about 11″ x 15″ (279 mm x 381 mm).

4

Fig. 3-16. Heat the plastic until it sags.

4 The amount of sag can be determined by experience. Light smoking of polystyrene is normal. Polyethylene and polypropylene sheet turn clear when they are ready to form.

5

Fig. 3-17. Bring plastic and mold together.

5 Either lower the plastic or raise the mold as required. **Important:** a **tight seal** between the **plastic** and the **mold** is needed.

6 Apply the vacuum, air pressure, or mechanical pressure as needed.

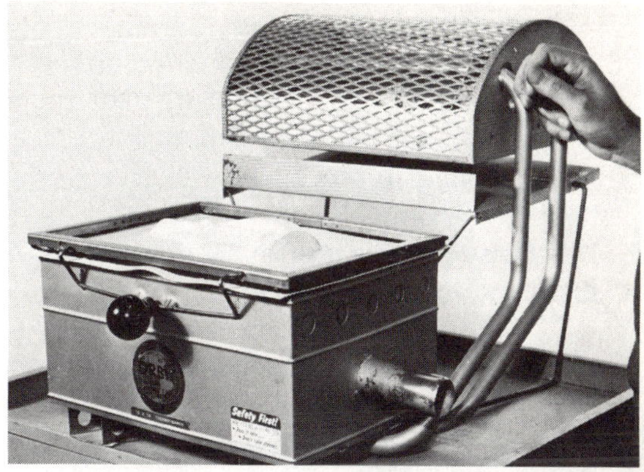 7

Fig. 3-18. Remove heat and cool.

7 (a) Remove the heat from the plastic and let it cool.
(b) Use cooling fan or compressed air to cool the plastic (optional).

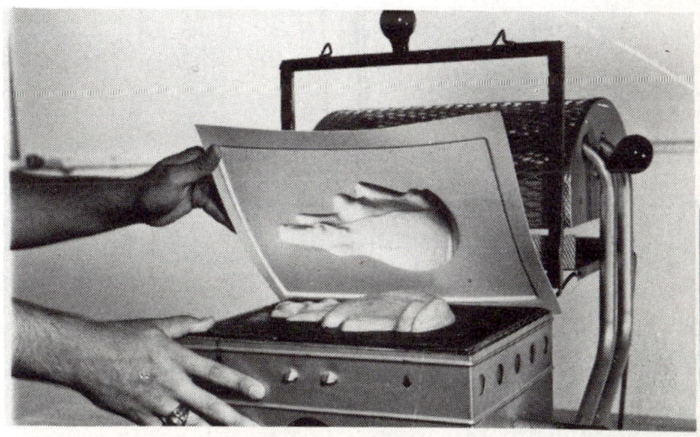 8

Fig. 3-19. Remove the plastic part from the mold.

8 Move the mold and plastic apart. Open the clamp frame. Take the plastic off the mold. Remove the plastic sheet from the clamping frame.

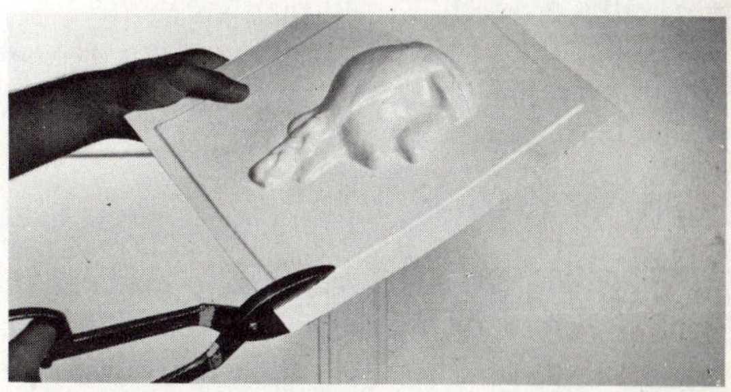 9

Fig. 3-20. Trim the part.

9 Trim the extra plastic from the part. Use a tin snips, paper cutter, sharp scissors, or a die cutter. The type of trimmer will depend on the part, the materials used, and the equipment available.

Safety Notes

The heating unit on a thermoformer should **always** be protected. Guards around the unit will prevent burns. If it is not well guarded, place a HOT sign on the heating element.

Be sure the heating element is turned **off** when not in use.

Thick plastic sheet holds heat a long time. **Cool it before handling. Use gloves.**

STUDENT ACTIVITIES

1. Thermoform several different plastic materials on each mold.
2. Materials experimentation:
   a. Vary the kind of materials on a given mold. Discover the differences in moldability and technique.
   b. Vary the material thickness. Notice the changes in the product made.
3. Process experimentation:
   a. Vary the heating time. Notice the changes in product quality. Notice the different sag in the plastic.
   b. Vary the heating speed by placing the heater at different distances from the material. Notice the changes in the product.
   c. Vary the speed of advance of the mold or plastic to the mold. Notice any change in the product.
   d. Vary the length of vacuum time. Notice what happens.

LAMINATING

Lamination is a group of processes in which layers of materials are formed into sheets or parts. Often they are pressed together under pressure. Lamination is divided into two groups. These are:
1. Low-pressure lamination and
2. High-pressure lamination.

Low-pressure laminates are formed at from 0 to 1000 psi (7000 kPa). Pressures over 1000 psi form high-pressure laminates. Two types of low-pressure lamination are discussed in this chapter. They are (1) fiberglass reinforced plastics and (2) thermoplastic laminates.

Reinforced Plastics — Process Description

Reinforcing is a lamination process in which fibers are added to plastic materials to form a stronger product. Glass fibers are usually the reinforcing fibers. Polyester plastic resin is the most common plastic component. Together, they form what is known as fiberglass reinforced plastics (FRP). The best qualities of the two are in the product made — the strength of the glass and the rigidity of the plastic.

Glass fibers can be combined with many thermoplastic and thermosetting plastics. Also, fibers other than glass may be used as rein-

Dake Corp.

Fig. 3-21. A 100-ton (90 Mt) molding press. It produces fiberglass reinforced tote boxes.

forcements. Boron filament and nylon strands are used in some cases.

Glass-filled thermoplastic pellets may be injection molded into reinforced products. Parts such as automobile fans and under-fender liners are often reinforced.

Glass fibers are also mixed with many thermosetting plastics during manufacturing. Some of these are phenolic, melamine, urea, and DAP materials. They are compression or transfer molded into many products.

Fiberglass Reinforced Plastic Lamination

Fiberglass reinforced laminates can be made. Glass fibers are added to the liquid thermosetting plastic in or on the mold. They may be combined in two ways. One is by hand lay-up. In this method, glass materials and polyester resin are put on the mold by hand. Hand lay-up is used in the school laboratory. It is also used in industry. The other method is machine lay-up.

Importance, Advantages, and Disadvantages of Fiberglass Reinforced Plastics

Fiberglass reinforced plastics make strong products. They are much stronger than many plastics without this reinforcement. The products are very durable. They are resistant to weather, chemicals, and electricity. Usually, they require less expensive molds than plastics in other processes. There seems to be no limit to the size of the fiberglass reinforced plastic parts. See Fig. 3-22. Color can be permanently molded into fiberglass products. A smooth, hard, shiny surface is possible.

Production of fiberglass products may be slower than other plastic processing methods. Using a number of molds will speed up production of large parts. Small parts are produced faster by compression molding. Fiberglass materials are messy to handle. They present a cleanup problem in the plant.

Industrial Equipment

A **fiberglass chopper and spray gun** combination is widely used. See Fig. 3-23. It costs less than most other fiberglass production equipment. It meters and sprays the right amount of plastic resin and catalyst. At the same time, it chops and sprays the glass fibers. Resin, catalyst, and fibers all meet on the mold at one time.

Commercial & Industrial Photography

Fig. 3-22. The hull of this fishing boat is fiberglass.

Venus Products, Inc.

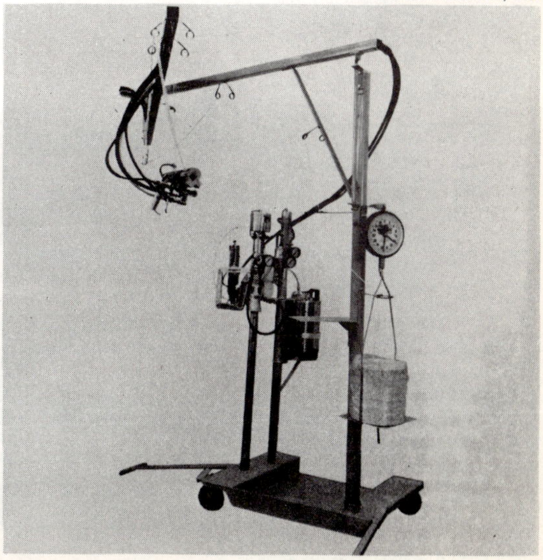

Fig. 3-23. Fiberglass chopper outfit.

McClean - Anderson, Inc.

Fig. 3-24. Laboratory FRP filament winding machine.

Process Equipment Corp.

Fig. 3-25. Filament wound tanks.

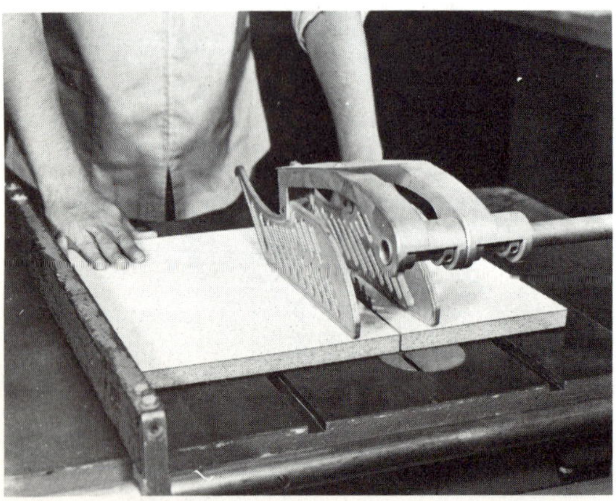

Fig. 3-26. Cutting Formica® molds for flat laminated FRP products.

The **filament winding** method produces super strong fiberglass products. Some of these are space rocket fuel tanks, pipe, and barrels.

Preforming is a method of molding formed fiberglass mats. The mats are made in a plenum chamber. Glass fibers and a resin binder are sprayed over a screenlike form. A fan pulls air through the screen. The glass fibers take the shape of the screen form. The preform is often put in an oven. This dries the resin binder which holds the fibers together. The preform is then placed in a compression mold. Resin is added. It is cured under heat and pressure.

Fiberglass Laminating in the Laboratory

Fiberglass laminating can be done in the school laboratory. Much of it is done by hand lay-up contact molding. Simple but useful contact molded products are hot pads, trivets, and clipboards.

A flat mold for these products is simple to make. Cut one or two pieces of high-pressure laminate (Formica). A sink top cutout is one source, Fig. 3-26.

Coat the Formica mold surfaces with carnauba paste wax. (Trewax® is one.) Apply several coats. Polish surfaces lightly to a high luster with a soft clean cloth.

Other molds for fiberglass may be made. They may also be purchased. Molds are available for such products as tables, cycle helmets, and trays. For more information about fiberglass work and mold making, read **Fiberglass: Projects and Procedures.**[1]

Safety Note

Use polyester resin only in a **well ventilated** room.

1. Authored by Gerald L. Steele and published by McKnight Publishing Company, 1962.

Checklist of Materials and Equipment Needed for Fiberglass Laminating

Assemble the following materials and equipment for fiberglass laminating:

1. Polyester resin.
   a. Gel coat (air-inhibited).
   b. Laminating resin (non-air-inhibited).
2. Catalyst.
3. Paper cups or other disposable measuring containers.
4. Color pigments (optional).
5. Eye droppers for measuring catalyst.
6. Mixing sticks or coffee stirrers.
7. Molds.
8. Mold release or carnauba wax. (Trewax® is one kind.)
9. Acetone or lacquer thinner for cleanup.
10. Cheap natural bristle brushes. (Soak the paint off the handles with acetone first.)
11. Plenty of wiping rags.
12. Several cans to clean brushes in.
13. Bench covers.
14. Protective clothes.
15. Plastic gloves (disposable).
16. Glass mat, cloth, or roving.
17. Styrene monomer to thin resin, if needed.

Refrigerate polyester resins and MEK catalysts for longer shelf life.

Safety Note

Do **not** store MEK peroxide at a temperature below 40° F (5° C). It may crystallize.

Do **not** store food in the same refrigerator. It will cause the food to taste of the peroxide. This may cause **illness.**

Operation Sequence for Fiberglass Laminating

There are five steps in fiberglass laminating. These are:

1. Get the mold(s) ready.
2. Measure, mix, apply, and cure the gel coat.
3. Measure, mix, and apply the polyester resin and glass materials.
4. Cure the laminate and remove it from the mold.
5. Trim the laminate.

The molding sequence is shown in the following illustrations. **Read the resin manufacturer's instructions very carefully. Check the amount of catalyst recommended.** Make sure to cover the benchtops before starting. Wear protective clothing. Wear disposable plastic gloves that will protect the hands from the plastic resin and glass fibers. Assemble all needed materials before starting.

1 **Prepare the mold(s):** Apply mold release to the mold.
(a) Paste type — rub on and polish out two coats. Use a soft cloth.
(b) Film forming (PVA) — brush, dip, spray, or rub on a thin coat.
Allow to dry completely.

1

Fig. 3-27. Putting on the mold release.

2 **Apply the gel coat:**

(a) Measure the gel coat resin. It will take about one fluid ounce per square foot (0.03 ml per cm^2).

(b) Add a color pigment **only if a solid color is wanted.**

(c) Measure and mix in 8 to 12 drops of MEK peroxide catalyst **per ounce** (0.27 to 0.40 drops per ml or 1 drop/4 ml to 1 drop/2.5 ml) of resin.
Mix well. If cobalt promotor must be added, use a separate eye dropper.
CAUTION: Do not mix cobalt and catalyst directly together. An explosion may result.

(d) Brush on the gel coat to an even 1/32″ (1 mm) layer (about the thickness of tablet-back cardboard). Allow the gel coat to **cure** about 1 to 1-1/2 hours.

(e) **Clean the brush in acetone or lacquer thinner.**

2

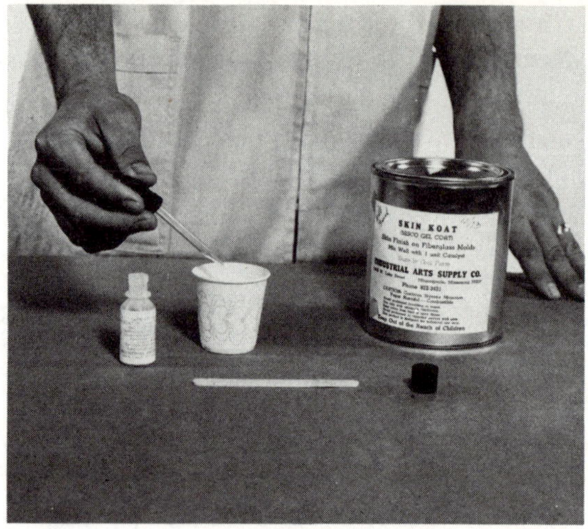

Fig. 3-28. Mixing the catalyst for the gel coat.

3A

Fig. 3-29. Laying on the decoration.

3B

Fig. 3-30. Saturating the mat with resin.

3A Apply the lamination layer:

(a) Measure out 8 ounces (0.25 ml per cm^2) of laminating resin per square foot of mold area. Mix in thoroughly 32 to 48 drops of MEK catalyst per 8 ounces (236 ml) resin or 4 to 6 drops per ounce (0.13 to 0.20 drops per ml or 1 drop/8 ml to 1 drop/5 ml) of resin.

(b) Brush a generous coat of resin over the cured gel coat. Add the decoration or glass materials to the fresh resin.

(c) Add decoration at this point (optional).
   (1) Decorative mat.
   (2) Cotton print cloth **face down.** Brush resin over the print cloth.
   (3) Metallic stars, metal flakes, or other decorative materials.
   (4) Color pigment in resin (optional).

3B Add fiberglass mat (or other glass material) at this point. Lay them flat. Brush or stipple resin into the glass material until all air (whiteness) disappears. **Clean the brush.**

A second lamination layer may be added to the fresh resin or later after it is cured. **Removo any surfacing agent (wax)** if the first layer is cured. Glass cloth, mat, or chopped fibers may be used for this and/or the other lamination layers. **Clean the brush.**

3C

Fig. 3-31. Double flat mold with cement block weight.

3C A second mold may be placed over the back of the lamination layers. It should have a cured gel coat on it, also. Weights can be put on it to compress the laminate or the molds may be put in a press.

4

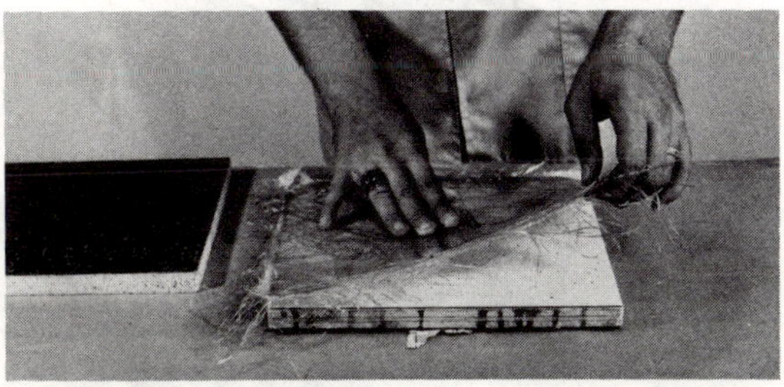

Fig. 3-32. Removing the part from the mold.

4 After the resin is cured, pry the part from the mold with a stick or give a few light taps with a **rubber mallet** to loosen the part.

5

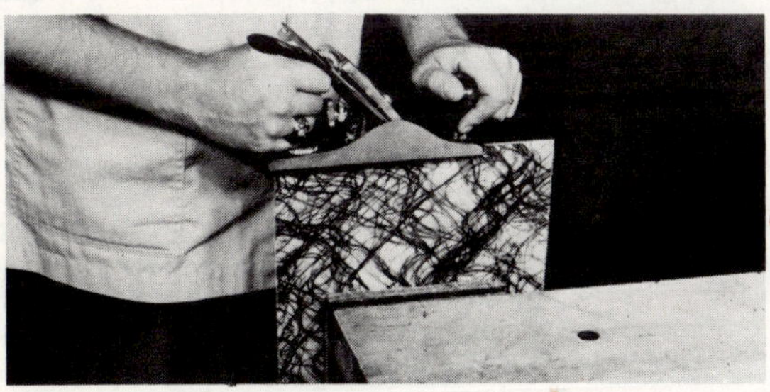

Fig. 3-33. Planing the edges.

5 The edges may be planed to get them straight. Trim the part with a jig saw, hacksaw, or metal cutting band saw. Tin snips can be used for thin laminates.

Thermoplastic Laminating – Process Description

Thermoplastic lamination is a common process. A photograph, identification card, leaf specimen, or a similar flat item may be used. It is sandwiched between two or more thermoplastic sheets. The object and plastic sheets are combined under heat and pressure. A single product is formed. It is called a **laminate.**

Materials for Laboratory Thermoplastic Laminating

Most thermoplastic sheets will work for this process. Clear .010″ (0.25 mm) vinyl (polyvinyl chloride) or acetate (cellulose acetate) sheets are used most often. Dark colored sheets may be used for the background. A clear sheet is placed over the article. An article must meet two requirements to be laminated. It must be dry. It must be able to stand the heat and pressure of lamination. Any article meeting these requirements may be used in the "sandwich." A limited amount of plastic can be forced around an object in lamination. **The total thickness of the plastic sheets should be about three to four times the thickness of the object enclosed.**[2] More plastic may be used to make a thicker laminate.

Fig. 3-34. Assembly of a thermoplastic lamination sandwich.

2. Cherry, Raymond, General Plastics, Bloomington, Illinois: McKnight Publishing Company, 1967, p. 255.

Operation Sequence for Thermoplastic Laminating

There are six steps in thermoplastic laminating. These are:

1. Assemble the materials to form the sandwich(es).
2. Place the sandwich(es) between the heated press platens.
3. Apply a low platen pressure for a short time to soften the laminating plastic.
4. Apply a higher pressure for a long time to compress and bond the laminate(s) together.
5. Cool the laminate(s) under pressure.
6. Remove and trim the laminate(s).

The laminating sequence is shown in the following illustrations. **Read the manufacturer's instruction manual for exact specifications and instructions.**

1

Fig. 3-35. Thermoplastic lamination ready to put in press.

2

Fig. 3-36. Putting the thermoplastic lamination sandwich into the press.

1 (a) Close the press platens.
(b) Turn off the water flow.
(c) Turn on the press platen heater switch. Heat it up to 300° F (149° C).

While the press is heating, assemble the sandwich(es) for laminating. The following is an example of a single sandwich.

(Top)
2 cushion blotters.
1 chrome polish plate, face down.
1 or more plastic sheets.
(Center)
Article(s) to be laminated.
(Bottom)
1 or more plastic sheets.
1 chrome polish plate, face up.
2 cushion blotters.

Be sure the plastic sheets are at least 1/4″ (6 mm) larger **all around** than the article being laminated.

2 Open the press platens when the press reaches 300° F (149° C). Put the sandwich(es) between the two heated press platens. **Be careful. The press platens are hot. Use gloves.** Close the press platens on the sandwich(es).

3 Apply a low pressure of about 10 psi (0.703 kg/cm^2) to the sandwich for 1-1/2 to 16 minutes depending on the number of sandwiches being laminated at a time. See Fig. 3-39 for correct times.

Figure the **total ram pressure** this way: Length in inches (millimeters) x width in inches (millimeters) x 10 (0.703 = TOTAL RAM PRESSURE.

A 9″ x 12″ (229 mm x 305 mm) lamination will need about 1080 pounds (489 kg) total ram pressure for preheating.

4

Fig. 3-37. Applying the pressure after preheating.

6

Fig. 3-38. Releasing the thermoplastic sandwich by flexing.

4 Apply 300 psi (21.09 kg/cm^2) pressure for 2-1/2 to 20 minutes depending on the number of sandwiches in the lamination. See Fig. 3-39 for the exact time.

A 9″ x 12″ (229 mm x 305 mm) lamination will need about 32,400 pounds (14,696 kg or 14.7 Mt) **total ram pressure,** for bonding. If the pressure drops below this as it bonds, it is not necessary to add more pressure.

5 (a) After the bonding time is up, turn off the platen heater switch.

(b) Turn on the water valve **very slowly** and run the water through the platens to cool them.

6 (a) After about 6 minutes of cooling (longer for several sandwiches), the platens should drop to about 100° F (38° C) or 150° F (66° C). Open the platens and take out the laminate(s).

(b) Flex the laminate(s). It (They) will pop off from the polished plates.

(c) Close the platens and turn off the water. Turn on the platen switches to start another cycle.

Multiple sandwiches may be made up and pressed at the same time. One blotter pad and two polish plates (back to back) must be put between each sandwich. Follow Fig. 3-39 for heating and bonding times for multiple sandwich lamination.

300°F (148.8°C) Platen Temperature, 10 psi (0.703 kg/cm^2) preheat pressure and 300 psi (21.09 kg/cm^2) bonding pressure for a 9″ x 12″ (228.6 mm x 304.8 mm) sheet.

Number of Sandwiches	Preheat Time (Minutes)	Bonding Time (Minutes)
1	1½	2½
2	3½	5
3	5	8
4	8	11
5	13	15
6	16	20

Information furnished by the Dake Corporation, Grand Haven, Michigan.

Fig. 3-39. Time, temperature, and pressure needed for thermoplastic lamination of vinyl plastic sheets.
Information furnished by the Dake Corporation, Grand Haven, Michigan.

STUDENT ACTIVITIES

1. Make a clipboard, hot pad, or other flat fiberglass product on a Formica laminated mold.

2. Make products from fiberglass on purchased or laboratory-made molds.

3. Materials experimentation.
   a. Vary the resin types used.
   b. Vary the catalyst percentage on experimental pieces. **Caution: Do not exceed 4% catalyst concentrations.**
   c. Use different fiberglass reinforcing materials. Test the resulting products for strength and durability.
4. Process experimentation (vary a and b).
   a. Lay-up procedure:
      1) Spray-up.
      2) Hand lay-up of mat.
      3) Hand lay-up of cloth.
   b. Molding procedure:
      1) Contact molding.
      2) Compression molding.
5. Design products which can be made by fiberglass lamination.
   a. Flat shapes and profiles (long, continuous plastic shapes).
   b. Curved shapes and profiles.
   c. Irregular shapes and profiles.
6. Laminate photos, cards, biology specimens, coins, and similar objects between thermoplastic sheets.

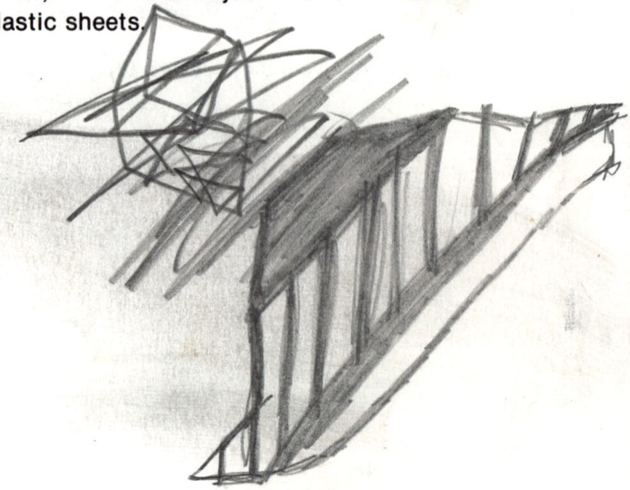

Chapter 4 CASTING AND FOAMING

Casting and foaming include several casting processes, foaming (expansion), and thermofusion.

In **casting processes,** plastic material is poured into the mold. Only gravity and atmospheric pressure are used. No other force is applied to the material in the mold. Plastic material may be in liquid or grain form. Casting processes use both thermoplastics and thermosetting plastics.

In **expansion processes,** plastic material is poured (cast) or forced (molded) into or onto a mold. They are then expanded. The expanded plastics are made up of cells. Both open and closed cell structures are formed. Plastics may be molded by injection, compression, and extrusion. They may be sprayed on or poured. Or they may be formed from expandable beads. Both thermoplastics and thermosetting plastics may be expanded.

Thermofusion processes form plastic materials by fusing (melting) them together. This is done on or in a mold or product. Thermofusion means heat fusion. Only gravity and atmospheric pressure are used. No other force is applied to the plastic during processing. Plastic powders, pellets, and vinyl plastisols may be used. Thermofusion includes several molding, casting, and coating processes. These may be **static** (at rest) or **dynamic** (in motion). Vinyl dip casting and coating and stationary powder molding are examples. Slush casting and rotational molding are also included.

Equipment for casting and foaming may be very simple or very complex. It varies depending on many factors. Some processes are very slow and require much hand labor. Others are highly automated. A wide variety of materials is used.

CASTING PLASTIC MATERIALS

Casting resins include clear casting resins and water-extended polyester (WEP) resins. Epoxy, silicone, DAP, urethane, acrylic, and phenolic casting resins are also used.

Importance, Advantages and Disadvantages of Casting

Casting equipment is very inexpensive. Sometimes many different types of plastic materials may be processed in the same molds. Molds do not have to stand high pressures. They may be made of cast or machined aluminum. Plastic, gypsum cement, silicone RTV, and urethane rubber compounds are also used. Even thermoformed polyethylene and high impact polystyrene molds have been used successfully.

Casting is important as a technique for making molds. It is especially good for silicone RTV, urethane, and mass casting epoxy resins.

Casting has one main disadvantage. It is often slower than injection, compression, and transfer molding. Molds frequently are filled by hand.

Casting Resins – Process Description

Thermoplastic and thermosetting casting resins are poured directly into a mold and cured. The are used for a number of products in industry. Probably the earliest use of plastic casting was the first Celluloid (thermoplastic) billiard ball. See Fig. 4-3. It was made by John Wesley Hyatt in about 1868. Dr. Leo H. Baekeland's phenolic (thermosetting) plastic was also cast into many items. Some phenolic parts are still being cast.

General Electric Company

Fig. 4-1. Silicone rubber mold used to produce fine detail castings.

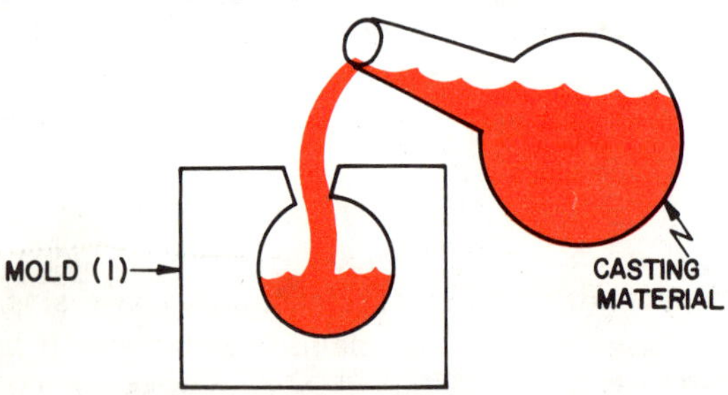

Fig. 4-2. In casting, material is poured into a mold by gravity.

Laboratory Casting of Polyester Resins

Castings of clear polyester resin and water-extended polyester (WEP) resin are included in this chapter.

Polyester resins are thermosetting plastics. They polymerize, or cure, by **exothermic heat** (heat generated and **given off**). The exothermic heat is caused by a chemical reaction between the promotor and the catalyst. The **promotor** is a reactive chemical. The catalyst is a chemical agent. The promotor is usually added to the polyester resin at the factory. It is often a cobalt metal dispersion. Cobalt is highly reactive in the presence of oxygen. Methyl-ethyl-ketone (MEK) peroxide contains about 11% free oxygen. It is used for a catalyst. The oxygen in the MEK peroxide catalyst combines with (oxidizes) the cobalt metal. Heat is created. This polymerizes (cures) the polyester resin.

Clear polyester resins may be cast in glass, silicone RTV, and flexible urethane. Latex rubber, thermoformed polyethylene, glazed ceramic, and polyester molds are also used. Glass jars and tumblers may be used for molds. Glass and thermoformed molds may be purchased. Handicraft and plastics supply houses usually stock them.

Celanese Plastics Company

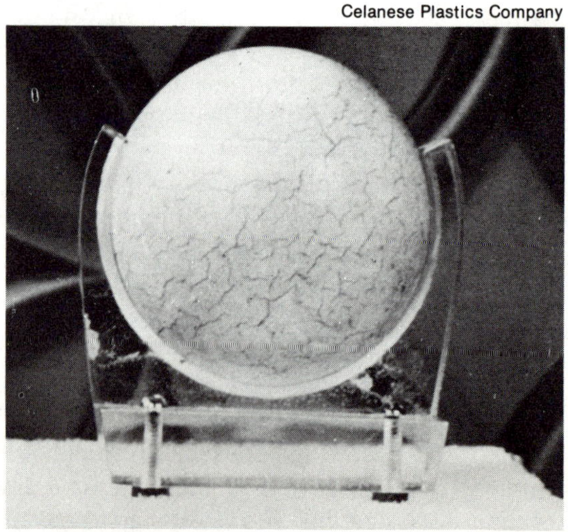

Fig. 4-3. Hyatt's Billiard Ball — Cast Cellulose Nitrate.

Polymer Corp.

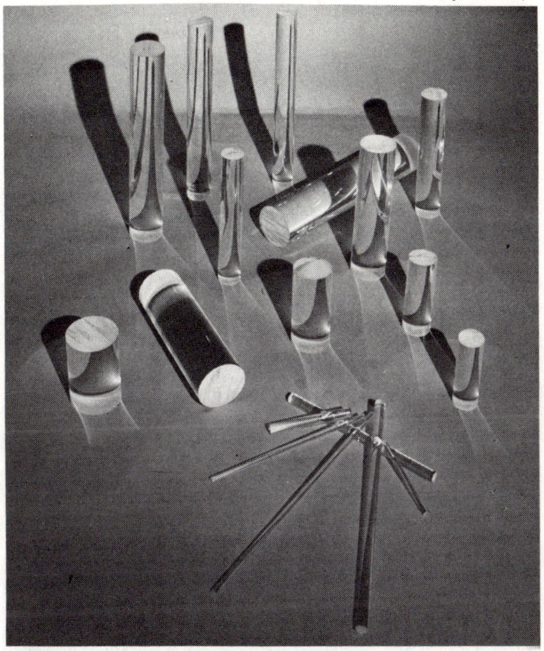

Fig. 4-4. Cast acrylic plastic rods

Polyester resins and catalysts last longer if refrigerated. Do not store MEK peroxide at less than 40° F (5° C). It may crystalize.

Safety Note

Do **not** store food in the same refrigerator with the resins and catalysts. They may cause the food to be spoiled.

When using polyester resins, throw away used paper cups and mixing sticks. Destroy sticky bench covers when finished. Clean your hands carefully. Check with your instructor for a safe cleaning solution.

Operation Sequence for Polyester Casting Resins

There are five steps in the process of casting with clear polyester resins. These are:

1. Select and prepare the mold.
2. Measure the amount of resin needed.
3. Add the right amount of catalyst and mix well.
4. Pour the material into the mold.
5. Cure the resin and remove it from the mold.

Safety Note

Do not directly mix catalyst and promotor together. An explosion may result.

Use the materials **only** in a **well ventilated** room.

Cover the bench tops to keep the resin from sticking. **Wear protective clothing, including plastic gloves.** Assemble all the materials needed before starting the process. The following illustrations show the operation sequence of polyester casting. **Read the resin manufacturer's instructions. Observe any precautions given.**

---

1 (a) Prepare the molds. Apply a thin coat of mold release to polyester, glazed ceramic, and reusable glass molds.
(b) Silicone RTV, flexible urethane, and vacuum-formed polyethylene molds usually do not need mold release.
(c) One-use glass molds do not need mold release.

2 Pour right amount of casting resin into a paper measuring cup. Pour only enough resin for immediate use. Mold volume can be measured with water. Pour the water out of the mold into a measuring cup to get the volume. Blow the mold dry with an air gun.

3A

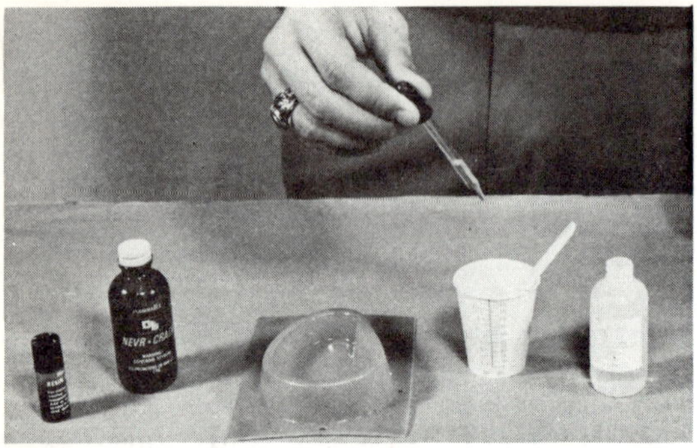

Fig. 4-5. Using an eyedropper to add catalyst to clear polyester casting resin.

3A Add the right amount of catalyst to resin/promotor mix. Use same measuring method as above. **Caution:** Thicker castings require less catalyst. **Consult the manufacturer's recommendations.** One manufacturer suggests the following:

Casting Thickness	Catalyst Concentration
1/8″ or less (3.1 mm or less)	16 drops per fluid oz. (0.54 drops/ml or 1 drop/2 ml)
1/8″ to 1/4″ (3.1 to 6.3 mm)	12 drops per oz. (0.41 drops/ml or 1 drop/2.5 ml)
1/4″ to 1/2″ (6.3 to 12.7 mm)	8 drops per oz. (0.27 drops/ml or 1 drop/4 ml)
1/2″ (12.7 mm) and over	4 drops per oz. (0.13 drops/ml or 1 drop/8 ml)

1 fluid oz. (29.5 ml) = 2 tablespoons (29.5 ml)

3B

Fig. 4-6. Mixing the catalyst with the resin.

3B Add color concentrate if desired. Stir in without mixing in air. **Caution: Do not mix promotor and catalyst directly together. Violent explosion may result.**

4A

Fig. 4-7. Filling a vacuum formed polyethylene mold with clear polyester casting resin.

4A Pour the resin into the mold. Fill the mold to the top if it is a one-layer casting. Allow to cure.

4B

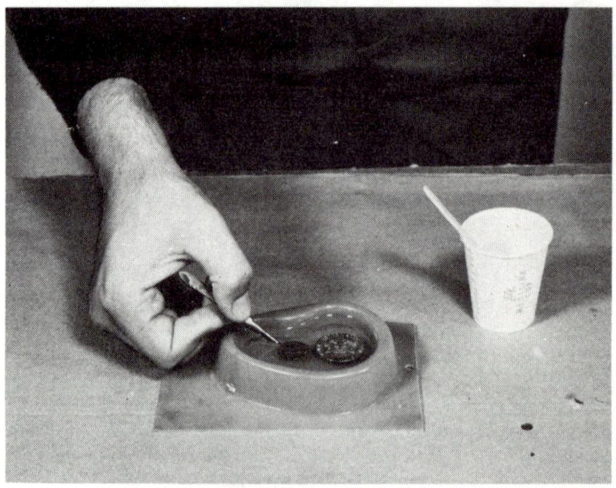

Fig. 4-8. Placing a coin in clear polyester casting resin.

4B (a) Fill the mold to the desired height for a multilayered casting. Cure first layer. Add a colored layer or the article to be embedded. "Wet" the coil or object to be embedded in the resin first.

(b) If wire inserts are used, place them in after plastic resin is poured. To prevent cracking around wires or other embedments, dip the wire or embedment in an anti-cracking solution.

5 Remove the product from the mold after curing. Curing may take 2 to 6 hours. Rigid glass or ceramic molds may be placed in warm water to help removal if necessary.

Cast polyester parts may be sanded and buffed, if necessary, to get a final finish. The acrylic plastic buffing procedure may be used.

1. Rough sand to remove deep marks.
2. Fine sand with progressively finer paper. Wet sand to 400 or 600 grit. Always sand at right angles to the last sanding when making each change to finer paper.

Ashland Chemical Company

Fig. 4-9. Cast water-extended polyester resin product.

3. Buff with 1725 rpm buffer with a 6″ to 8″ (152 mm to 203 mm) loose-sewn buffing wheel and acrylic buffing compound.

Water-Extended Polyester Casting Resins

Water-extended polyester (WEP) castings are used for lawn and garden ornaments. They may be used for decorative and useful indoor articles. WEP castings make sturdy, long-lasting thermoforming molds for laboratory use.

WEP resins are similar to regular polyester resins. There is one important difference. They are made so that water may be mixed in. The usual mixture is 50% water and 50% WEP resin. This mixture gives maximum strength at minimum cost. As much as 75% water may be added. No special mixing equipment is needed. Adding more water makes the mixture **thicker.** A thick mixture is harder to work with.

A cobalt-MEK peroxide promotor and catalyst system is used. The cobalt promotor is usually mixed in at the factory. The MEK peroxide comes in a separate container. It is added to the mixture **last.** The chemical reaction of this system is the same as for other polyester resins. When not in use, polyester resins and catalysts should be stored in a refrigerator. They will stay usable longer.

Safety Note

Do **not** store MEK peroxide below 40° F (5° C). It may crystalize.

Do **not** store food in the same refrigerator with polyester resins and catalysts. They may cause the food to be spoiled.

WEP emulsions use the same types of molds as used for clear polyester casting. Thermoformed polyethylene molds are the best. No mold release is required for them.

Operation Sequence for Water-Extended Polyester Casting

There are six steps in the process of water-extended polyester casting. These are:

1. Figure the amount of materials needed.
2. Weigh out the resin and water in **separate disposable containers.** Figure out the amount of catalyst needed.
3. Start up the high speed mixer and add the water **very slowly** into the vortex (center cone). Let the mixture stand to permit air to escape.
4. Add the catalyst and stir gently.
5. Pour the mixture slowly into the mold.
6. Allow the part to cure before removing it.

The following illustrations show the operation sequence for water-extended polyester casting. **Read the resin manufacturer's instructions carefully. Check the amount of catalyst recommended. Wear protective clothing, including plastic gloves.** Assemble all materials needed before starting the process.

Safety Note

Use polyester resin **only** in a **well ventilated** room.

1 Figure out the amount of materials needed.

(a) Decide on the polyester/water ratio. Try a 50/50 ratio to start.

(b) Weigh out, in a disposable paper container, water equal to about 55% of the volume of the mold to be used. Record this weight. Be sure to subtract the weight of the container.

2

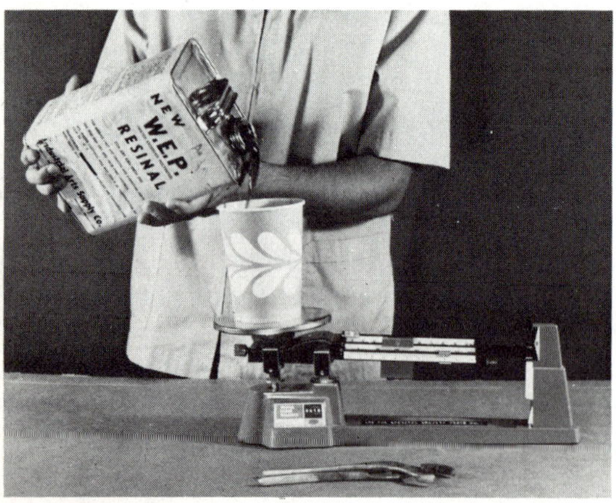

Fig. 4-10. Weighing resin.

2 Weigh out WEP resin equal in weight to the water just weighed. Be sure to subtract the weight of the container. Use a wide-mouthed container about 4 times the volume of the resin to be used.

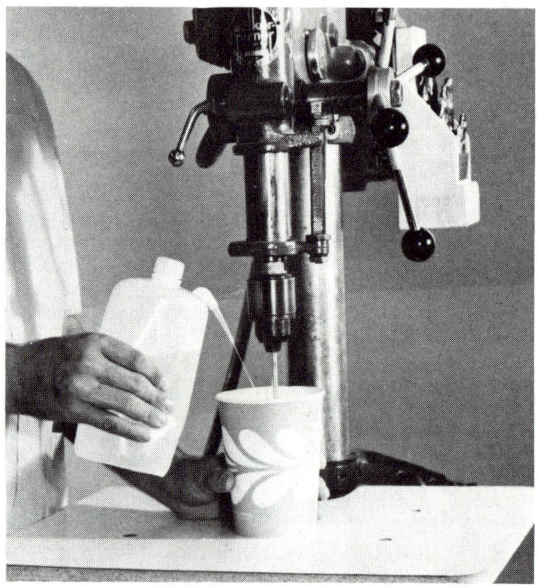 3

Fig. 4-11. Adding water, using a "wash" bottle.

3 (a) Put the mixer blade in the drill press or drill and set the speed at 2000 rpm.

(b) Put the weighed out water in a plastic "wash" bottle.

(c) Turn the mixer on and add the water **very slowly** to the vortex of the resin. Continue to mix for about a minute. A polyester color pigment may be added at this point.

(d) Clean the mixing blade with acetone when work is completed.

4

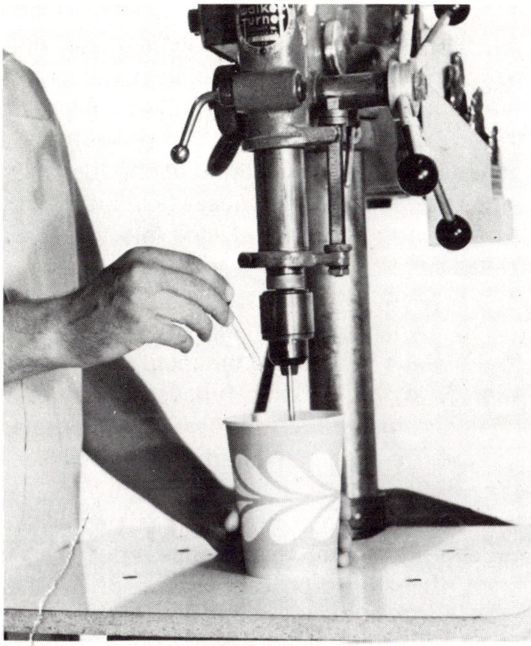

Fig. 4-12. Adding catalyst to WEP resin.

4 (a) Let the mixture stand for several minutes to 24 hours to let the air escape.

(b) Figure out the right amount of catalyst to be used. From 1/2% to 3/4% (0.5% to 0.75%) based on WEP resin weight is usually recommended.

Caution: Do not mix catalyst and promotor directly together.

(c) Add the catalyst to the resin/water mixture with an eye dropper. Stir the catalyst in with the mixer blade without mixing air into the emulsion.

NOTE: One-half of one percent catalyst (0.5%) is about one (1) drop catalyst per fluid ounce (0.03 drops/ml) of resin.

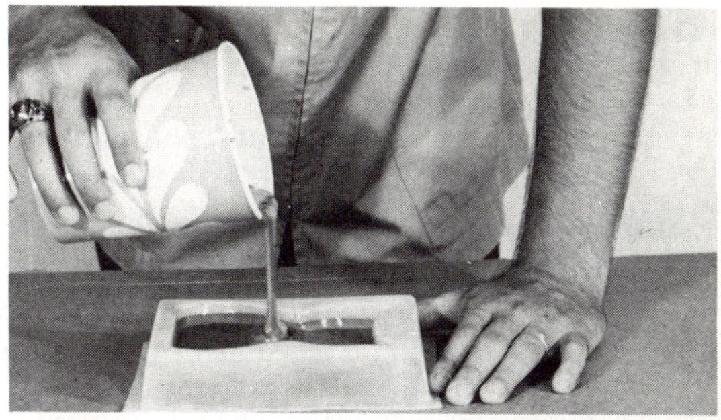 5

Fig. 4-13. Filling the mold with WEP resin.

5 (a) Slowly pour the mixture into the mold. Pour it to fill the difficult areas first. Brush some emulsion on fine detailed areas.

(b) Vibrate or flex the mold to get out the trapped air.

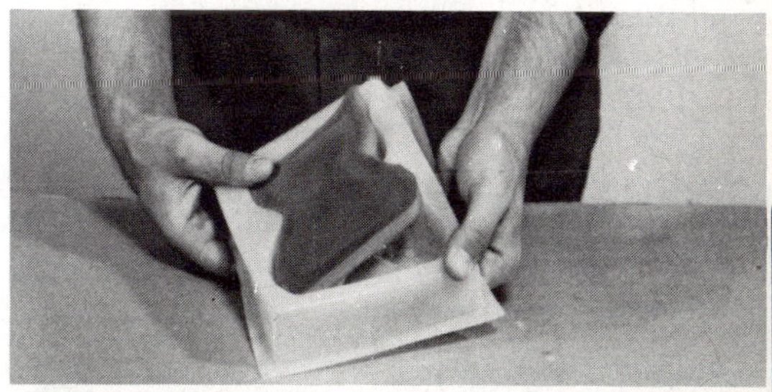 6

Fig. 4-14. Removing cured WEP resin from the mold.

6 (a) Let the emulsion cure. A 50/50 mixture should solidify in about 5 minutes. The casting should be about 90% cured in 1 hour.

(b) The casting may be taken out of the mold 10 to 20 minutes after pouring, if it is hard.

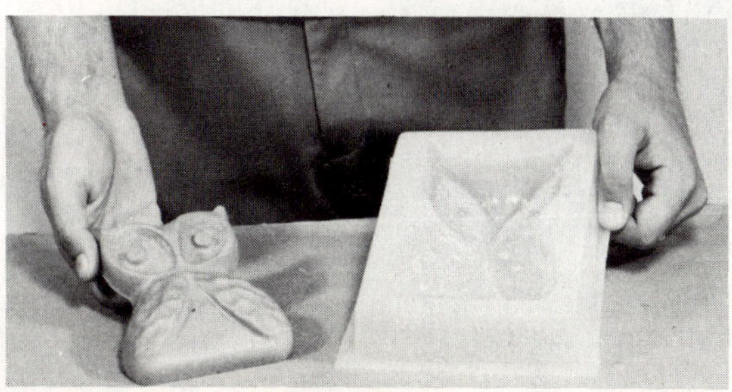

Fig. 4-15. Completed WEP casting taken from the mold.

STUDENT ACTIVITIES

1. Materials experimentation.
   a. Use various times, colors, and densities of materials.
   b. Use different proportions of water in WEP resins. The base 50/50 ratio may be varied to 75% water/25% resin and vice versa. Bake the water out of cured WEP products (150° F or 65 ° C) for a lighter product.
2. Process experimentation.
   a. Vary catalyst concentrations **slightly** in casting resins. Observe the variation in polymerization times.
   b. Experiment with different mold materials for each casting process.
3. Design products and molds for each casting process.
4. Make molds for casting by thermoforming, rubber casting, and metal machining.

THERMOFUSION

Process Description

Thermofusion means **heat fusion.** It is a name given to a group of similar plastics processes. It is the melting and sticking together of plastics particles using heat. In these processes, plastics are fused (melted) together in or on a mold surface. Only gravity and atmospheric pressure are used. No other force is added to the plastic material during fusion.

There are several processes in the thermofusion group. They are divided into two types. These are:

1. Static (stationary) and
2. Dynamic (in motion).

In the **static** processes, the mold is stationary (static). In the **dynamic** processes, the mold rotates (turns). Either powdered or liquid plastics may be used in most of these processes.

Static powder molding, single-axis, and two-axis rotational molding are technically **casting** processes. The plastics industry calls them molding processes more often than casting. For this reason, they will be called **molding** processes in this chapter.

Fig. 4-16. Static molding of thermoplastic injection pellets on a cookie sheet coated with Teflon or in aluminum molds. Several completed products are shown in the foreground.

Thermofusion Materials

The plastic material used in thermofusion may be either dry powder or liquid. Almost any finely powdered plastic may be used. Presently, **finely ground** polyethylene is probably the most popular. Vinyl plastisol is used as a **liquid** thermofusion plastic.

THERMOFUSION	
Static (Stationary)	**Dynamic** (In Motion)
1. Dip casting and coating.	1. Single-axis rotational molding (Heiser® process).
2. Full mold casting.	2. Two-axis rotational molding.
3. Slush casting.	
4. Static powder molding (Engle® process).	

Dip Casting and Coating

In dip casting and coating, a metal mold or product is heated. At about 400° F (205° C), it is dipped into a cold (room temperature) liquid plastic. The heat melts the plastic touching the mold or product. A layer of plastic is built up (fused). The mold or product is then removed from the liquid plastic. The plastic-covered mold or product is put into an oven. The oven is set at about 300° F to 350° F (150° C to 175° C). The plastic is completely fused in the oven. It is also smoothed. The fused part is then cooled. If on a mold the fused plastisol may be removed when cool. Using a mold, the process is called **dip casting.** The fused plastisol may be left on a product permanently. This process is called **dip coating.**

Materials Used

Dip casting is usually done with vinyl plastisols. Various hardnesses may be used. This depends on the product to be made. Hardness is controlled by the amount of plasticizer added to the vinyl.

Importance, Advantages, and Disadvantages of the Process

Dip casting is a minor process in the plastics industry today. Usually only small parts are made by this method.

Flexible parts with thin walls may be made by dip casting. Examples are plastic gloves and stretch boots. Parts with thicker walls are also made by this process. Bicycle handle grips and coin purses are examples of these.

Dip casting is a fairly slow process. Long mold heating times are needed. Parts require time to cure. The evenness of part thickness depends on the mold heating and dipping time.

Industrial Equipment

Production lines for automatic dip casting have been developed. They include five sta-

McNeil - Femco - McNeil

Fig. 4-18. Automated dip casting line.

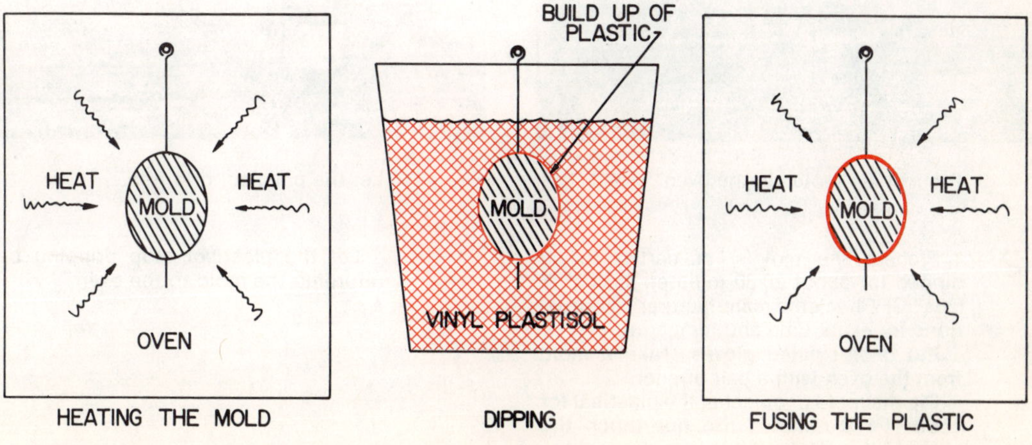

Fig. 4-17. Dip casting.

tions. These are a preheat oven, dip tank, fusing oven, cooling station, and a stripping station. The molds or parts to be dip cast are carried through this line on a conveyor belt. They are stripped (removed) from the mold by compressed air.

Operation Sequence for Dip Casting and Coating

Many interesting objects can be dip cast or coated in the school laboratory. Molds may be bought or easily made. Coin purses, key cases, fishing fly cases, and similar objects may be cast. Tool handles and other metal parts may be dip coated. Dip coating gives a smooth plastic surface.

Study the operation sequence in the following illustrations. **Read the labels on the cans of plastisol. Read any other instructions by the manufacturer.** Check the mold preheat temperatures, preheat times, cure times, and cure temperatures. They are **not** the same for each plastisol used.

1 Add color pigments and mix. Remove the air bubbles from the plastisol by letting stand 24 hours, jiggling, or vacuum de-airing it.

2

Fig. 4-19. Preheat dip molds in the oven.

3

Fig. 4-20. Let the plastisol drip free.

2 Preheat the mold(s) or part(s) to be dipped for about 20-30 minutes at 400° F (204° C). Check the manufacturer's instructions for exact time and temperature.

Put on insulated gloves. Take a mold from the oven with a pair of pliers.

Dip the mold or object in the plastisol for about 1-1/2 minutes. **Do not touch the sides of the plastisol can with the mold.**

3 Let the plastisol stop dripping before returning the mold to the oven.

4

Fig. 4-21. Put the dipped mold back in the oven to cure.

5

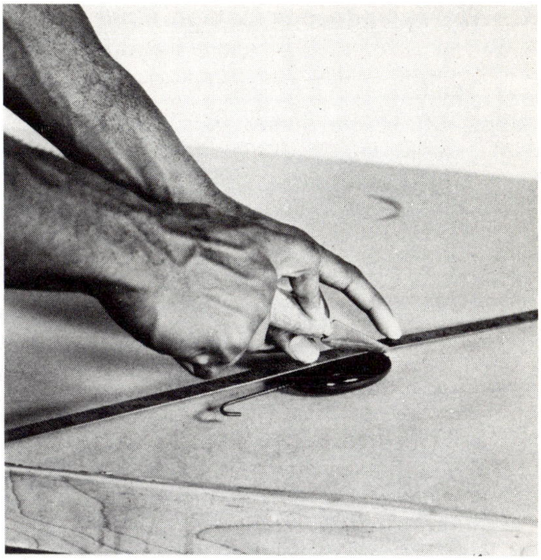

Fig. 4-22. Slit the coin purse from end to end.

4 Cure (fuse) the plastisol about 15 minutes at about 350° F (177° C). Check manufacturer's instructions for exact time and temperature.

Put on insulated gloves. Take the mold from the oven with a pair of pliers.

Put the cured part and mold in cool water. Cool it for at least 10 minutes. Remove from the water.

5 (a) Cut around the wire ends with a sharp knife.

(b) Cut a slit from end to end with a straight edge and a sharp knife.

(c) Carefully slip the part off the mold.

Full Mold Plastisol Casting

Vinyl plastisols cast in open molds form solid objects. Vinyl full mold casting has many of the same limitations, advantages, and disadvantages as dip casting. It should not be used for very thick castings. The process is usually not automated. Dip casting and full mold casting are done in different types of molds. That is the main difference between them. The plastisol is cast over a dip mold in dip casting. It is cast into a full mold in full mold casting. An example of full mold plastisol casting is bait casting.

Operation Sequence for Full Mold Bait Casting

The full mold bait casting process has five steps. Study the operation sequence in the following illustrations.

1

Fig. 4-23. Pour the bait plastisol into a cold mold cavity.

1 The mold should be at room temperature. Do not overfill the cavities. Plastisol expands when it heats. Fish hooks may be cast into the plastisol if they are wanted.

2

Fig. 4-24. Put the filled mold in the oven using pliers and gloves.

2 Cure them at 325-350° F (163° to 177° C) until the plastisol changes color completely.

Put on insulated gloves. Take the mold(s) from the oven. Put it in water to cool.

 3

Fig. 4-25. Remove the part from the mold.

3 Use a sharp knife to get under the edge of the plastic to get it loose. Pull from one end. Trim off any flash.

Slush Casting

Slush casting plastisols is similar to full mold plastisol casting. The difference is that a heated, hollow, open mold is used.

The vinyl plastisol is poured into the heated mold. It is allowed to solidify on the mold walls. When the desired wall thickness is reached, the extra plastisol is poured back into the container. It may be reused. The mold is put in the oven to fuse. The mold is taken from the oven and cooled. The part may then be removed.

Two-Axis Rotational Molding

Two-axis rotational molding or casting is a process for making completely closed hollow products. Balls, hobby horses, and tanks are made in this manner. The parts may be either rigid or flexible.

A hollow, two-part mold is used to make the product. Plastic powder or plastisol is placed inside. The mold is filled about 1/3 to 1/2 full. It is then closed. The mold is placed in an oven heated from 400° F to 700° F (205° C to 370° C). Then the mold is rotated in two directions at once. It turns over as it spins. As the mold rotates, the heat melts and fuses the plastic material. An even layer is formed inside the mold. Heating continues until all the plastic is fused. It is cooled by air and water spray as the mold continues to turn. When cool the mold can be opened and the part removed.

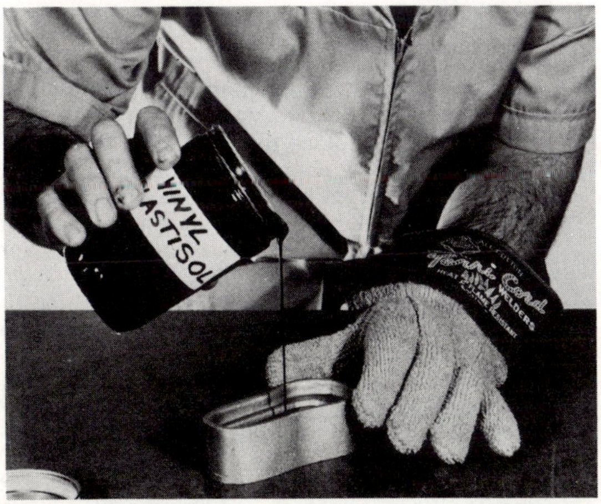

Fig. 4-26. Slush casting a vinyl plastisol door stop.

Materials Used in Industry

Many powdered thermoplastic materials may be used. Some of these are polyethylene, ethylene-vinyl acetate, polystyrene, and polypropylene. Polyvinyl chloride, cellulosics, and nylon may also be used. Special grades of vinyl plastisols are available for rotational molding flexible parts.

Industrial Equipment

Industrial two-axis rotational molding equipment varies in size. They range from 12″ x 12″ to 1200″ x 3600″ (300 mm x 300 mm to 30,000 mm x 92,000 mm). The overall dimensions of the maximum part size the machine can produce is used.

Fig. 4-27. Rotomolded table and chairs.

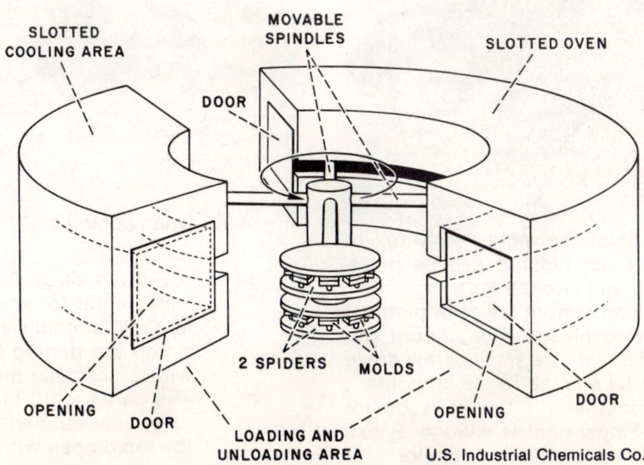

U.S. Industrial Chemicals Co.

Fig. 4-28. Four-spindle rotational molding machine.

Operation Sequence for Two-Axis Rotational Molding

The operation sequence for rotational molding is shown in the following illustrations.

1 Turn on the rotational molder oven. Warm it up to 450° F to 550° F (232° C to 288° C). Blow the mold dry with compressed air.

2

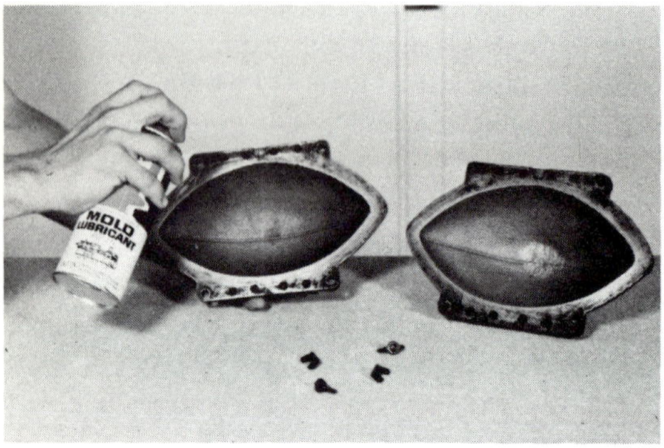

Fig. 4-29. Spray mold release.

2 Spray the inside of the mold halves with fluorocarbon mold release. Polish lightly with a soft cloth.

3

Fig. 4-30. Fill the mold.

3 Fill one half of the mold nearly full of plastic. The exact amount can be determined by trial and error.

Add a **small amount** of dry color pigment to the powdered plastic. The amount you can get on the end of a small screw driver blade should be enough for most molds.

Fasten it together. If it has bolts, do not overtighten. Finger-tight is enough. Shake the mold to mix the plastic and color.

4

Fig. 4-31. Put the mold in the oven and lock it in the arm.

4 Put on the insulated gloves. Open the rotational molder oven and put the mold in the mold holder. Make sure the mold is held tight. Close the door right away.

Turn on the rotation molder motor. Set it for a slow speed for large molds and faster for smaller molds. Set the timer for about 10 to 15 minutes. Use longer times for thicker or larger parts.

Put on insulated gloves. Turn off motor when the cure time is over. Take the mold out and air-cool it for 2 to 4 minutes, turning it in 2 directions by hand. **Be careful not to get burned.** Use a fan or low-pressure air jet (30 psi or 207 kPa) to help cool it.

5

Fig. 4-32. Open cooled mold and remove the part.

5 Put the air-cooled mold in a pail of luke warm water for about 5 minutes. Then put in a pail of cold water for 5 minutes more.

Tap the parting line of the mold lightly with a soft-face mallet to open the mold. The part should fall out. Use compressed air to blow out parts which stick. **Do not pry the mold open with a screw driver.**

STUDENT ACTIVITIES

1. Dip cast products on prepared molds in vinyl plastisol.
2. Again dip cast products on prepared molds. Use more than one color of vinyl plastisol to get a layered effect.
3. Experiment to find the best dipping time for a given plastic at a given mold temperature. Preheat several molds. Dip each one for a different length of time. Cure each and cool. Remove from the molds and compare them. Determine which dipping time makes the best part.
4. Dip coat tool handles with tool-dip plastisol. Coat the handles with the proper tool-dip primer first. Follow the procedure for dip coating. Do not strip the coating after it is cured.
5. Cast fishing bait and other lures with bait casting plastisol. Mix colors to make your product lifelike.
6. Slush cast door stops, football tees, and similar items.
7. Rotationally mold either powdered polyethylene or vinyl plastisol into products of many shapes. Use color pigments and metallic flakes to get varied decorative effects.

FOAMING PLASTICS

In expansion or foaming processes, plastic material is expanded in a mold. The expanded plastics are usually called foams. They have a cellular (spongelike) structure. Expanded plastics may be molded by extrusion, injection, and compression. They may be sprayed on or poured. They may be expanded from beads.

Plastic foams are used in many ways. Flotation, insulation, cushioning, impact absorbing, and packaging are a few uses. Nearly all plastics, both thermoplastic and thermosetting, may be expanded (foamed).

Two plastics are used most often to make plastic foams. They are polystyrene and polyurethane. Polystyrene (Styrofoam® and expandable polystyrene beads) is a thermoplastic. Polyurethane is a thermosetting plastic. Other common plastic foams are made from polyethylene, cellulose acetate, phenolic, silicone, and epoxy.

Both rigid and flexible foams are produced, Fig. 4-33. Rigid foams are used for boat flota-

Ransburg Corp. Scott Paper Co. Foam Div.

A. Rigid foam objects. B. Flexible foam speaker grilles.

Fig. 4-33. Products of resin foam casting.

Scott Paper Co. Foam Div.

Fig. 4-34. Flexible, open-cell urethane foam in gas tank to reduce a possibility of explosion.

Arco/Polymers, Inc.

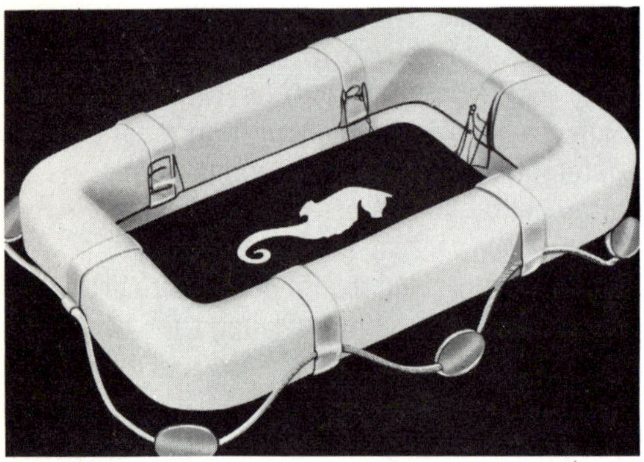

Fig. 4-35. Swim toy made from bead foam. A closed cell structure.

Fig. 4-36. Thermoformed duck decoy filled with urethane foam. Using the foam-in-place method.

tion cells, refrigerator insulation, and building insulation panels. Furniture frames can also be made of very dense rigid foams. Flexible foams are used for sponges, furniture padding, padded dash panels for cars, and mattresses.

Characteristics

Foams may have an open- or a closed-cell structure. An **open-cell structure** has openings between cells. Gases and liquids can move from cell to cell. A sponge has an open-celled structure. Foams with **closed-cell structure** have separate cell walls. Gases and liquids cannot move from cell to cell easily. Rigid foams usually have closed-cell structures. Flexible foams usually have open-cell structures.

Foam products are made in many densities. Density is the weight per unit of volume. A small, but heavy object would be high in density. Most plastic foams are **low** in density (lightweight). They range from 1/10 to 70 pounds per cubic foot (0.002 to 1.2 g/cm^3). Low densities provide good insulation and flotation characteristics. High densities make stronger, more rigid products.

Importance, Advantages, and Disadvantages of Foams

Foamed products are common. They are found in cars, the home, and industry.

Foams conserve plastic materials. Larger parts can be made with the same amount of raw plastic. A variety of techniques can be used to process foam. Equipment costs depend on the process used.

Many foamed plastics cure or solidify slowly in the mold. Slow curing tends to lengthen the cycle time.

Industrial Equipment

Foams may be processed by extrusion, compression molding, or injection molding. They may also be processed by spraying, pouring, or bead expansion. Special foam molding equipment may be built. Conventional extruders or injection molders are sometimes adapted for foam molding.

Resin Foam Casting

Urethane resin foams are thermosetting plastics. They polymerize by exothermic heat. The two parts of the resin mix are kept in separate containers. When mixed together, a chemical reaction results. The chemical reaction makes heat. Heat expands the gas and cures the foam.

Molds for resin foam casting may be of several types. Flexible silicone RTV and urethane rubber work well. Thermoformed polyethylene and high impact polystyrene are also usable. The resin foams release from the silicone RTV, urethane, and polyethylene. They stick to polystyrene. Plaster, gypsum cement, fiberglass, and cast aluminum molds have also been used successfully. These molds require several coats of a mold release.

Another way to cast resin foam is called the foam-in-place method. The foam is poured into an object. The object becomes a permanent mold. Examples of this are foamed-in-place insulation and boat flotation chambers. Duck decoys and other products can be made in the school laboratory. The outer shell can be thermoformed in two halves. The halves are glued together. Foam is put in from the bottom. See Fig. 4-36.

Operation Sequence for Resin Foam Casting

There are six steps in casting resin foam.

1. Prepare the mold.
2. Figure the volume of the mold.
3. Measure the resin components.
4. Add the resin components together and mix well.
5. Pour the mixture into the mold.
6. Cure the part and remove it from the mold.

Figures 4-37 through 4-40 show the operation sequence for resin foam casting. **Read the resin manufacturer's directions before starting. Observe any precautions given.** Keep all foam components in a cool, dry place. Many need to be stored and used at less than 75° F (25° C). Flexible foams absorb moisture. They should not be refrigerated. All foam materials should be capped tightly after each use. Clean the lid and top of the can well before closing. Clean the mixer blade with acetone or lacquer thinner. **CAUTION: Lacquer thinner and acetone are flammable. Wear protective clothing.**

Safety Note

Use resin foams only in a well ventilated room. They should be placed directly under a ventilating hood when foaming.

1

Fig. 4-37. Applying barrier coat to silicone mold.

1 Prepare the mold.

(a) Silicone RTV and urethane rubber molds: Mold release is often not needed. A barrier coat may be applied as a release agent. Spray, brush, or wipe it on.
(b) Vacuum-formed polyethylene: No release needed.
(c) Vacuum-formed high impact polystyrene: No release needed for flexible foams.
(d) Fiberglass and cast aluminum molds: Use fluorocarbon mold release.
(e) Plaster molds: Coat well with vaseline.

2/3

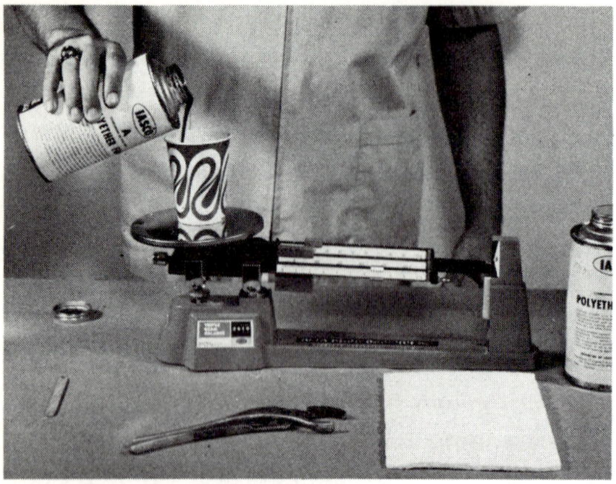

Fig. 4-38. Weighing urethane resin foam.

2/3 Figure and measure the foam components. Weigh each component separately. Add the weight of one to the next. For 50/50% mixtures, add equal quantities of each component.

4 Mix the two parts of the resin very well for about 20 to 30 seconds with a high shear paint mixing blade on an electric drill. The resin may also be mixed by hand with a stick.

5

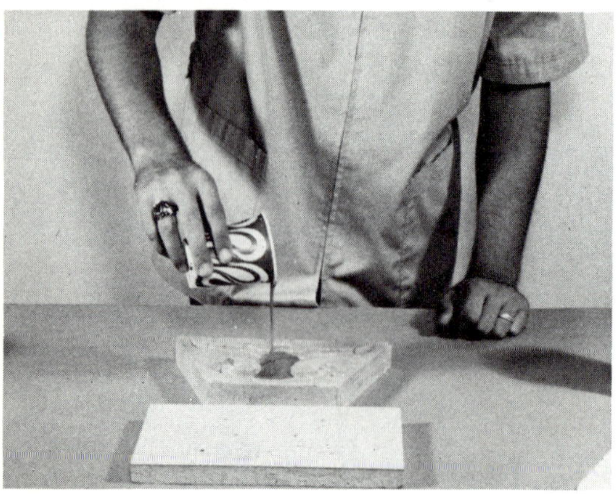

Fig. 4-39. Pouring resin foam.

5 (a) As soon as the mixing cup feels warm, stop mixing. The foam mix should be poured at once.

(b) Close the mold on the mix. A relief hole may need to be left open to let air and extra resin out. Use weights, clamps, or bolts to hold the mold closed.

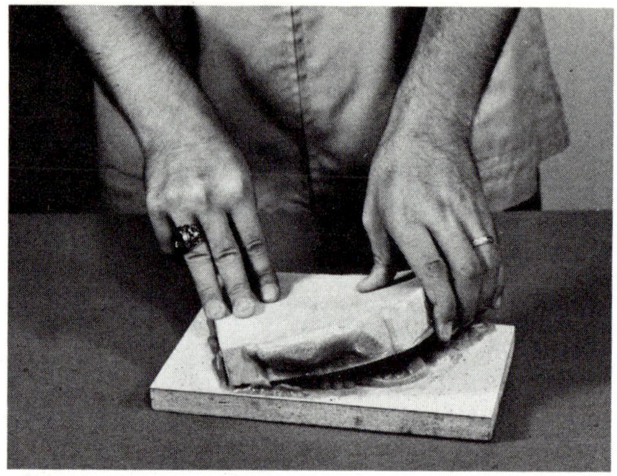 6

Fig. 4-40. Opening the mold after curing.

6 (a) Cure the product. Cure time of the foam depends on the resin used, mold temperatures, room temperature, and humidity. **Read the manufacturer's instructions for approximate times.** Some resins need oven cures. Follow the manufacturer's instructions **carefully** for this.

(b) Remove the part from the mold when the surface of the excess resin is tack-free. Open the mold. Trim off the flash with a scissors. Paint the surface if desired.

Expandable Polystyrene Bead Foam Molding

Expandable polystyrene bead foams are thermoplastics. They are small spheres with a sponge-like center. When the beads are heated, the cell walls soften and a gas inside expands. The softened beads expand and fuse together in a mold. They are then cooled. A solid closed cell mass results.

Virgin (unexpanded) polystyrene beads may be expanded up to about 50 times their original size. See Fig. 4-41. Bead foams are used for familiar things like picnic coolers, ice buckets, packaging cushions, and flotation devices.

Laboratory Bead Foam Molding

Bead foam molding may be done in the laboratory. Polystyrene beads are first pre-

Arco/Polymers, Inc.

Fig. 4-41. One pound of polystyrene bead foam before and after expansion.

expanded. Boiling water or radiant (dry heat) pre-expanders are used. Molding is usually done in aluminum molds. The molds are filled with pre-expanded beads. They are then put into a pressure cooker or a pail of boiling water. After processing for the required time, the molds are removed. They are cooled in a pail of cool water.

Checklist of Materials and Equipment Needed

Assemble the following materials and equipment for bead foam molding.

1. Electric stove or hot plate(s). Should be 230 volt, single phase, with 2000 to 2500 watt burners.
2. One or two five-gallon metal pails for pre-expanding beads.
3. A large pressure cooker (21-quart or larger).
4. Two or three five-gallon plastic pails for cooling water.
5. Assorted aluminum molds.
6. Silicone or fluorocarbon mold release.
7. Expandable polystyrene beads, white and/ or colored.
8. Dry color pigments for polystyrene beads (optional).
9. One or two large kitchen sieves to remove pre-expanded beads from water.
10. Two pairs of insulated gloves.
11. Drying frame or power drier for pre-expanded beads (optional).
12. Plastic storage containers for foam beads.

Arco/Polymers, Inc.

Fig. 4-42. Boat made of expandable polystyrene bead foam.

Operation Sequence for Bead Foam Molding

The process of bead foam molding has six steps.

1. Pre-expand the beads.
2. Prepare the mold(s).
3. Fill the mold(s) with pre-expanded beads and close it.
4. Fuse the beads in the mold under heat.
5. Cool the mold and the part inside.
6. Open the mold and remove the part.

The operation sequence is shown in the following illustrations. **Read the bead manufacturer's instructions.**

Safety Note

When handling the virgin polystyrene beads in large quantities, observe the following precautions.

a. A dangerous pentane gas gathers above large quantities of the beads. **A fire or explosion could occur.** Before removing the beads, open the bead barrel and allow the gas to escape.

b. **Use a paper or plastic container** to scoop out the beads. A metal cup might cause a spark of static electricity. That spark could ignite the gas. No apparent danger exists when the beads are packed in small plastic bags.

c. **Do not use virgin or pre-expanded beads near flame of any kind.**

1

Fig. 4-43. Stirring to get uniform expansion.

1 **Pre-expansion — hot water method:**

(a) Bring about 1 gallon (4 l) of water to boil in a 5-gallon (20 l) pail. Pour a few ounces of virgin beads into the boiling water. Stir until the proper bead density is obtained.

(b) Remove the pail from the burner. Dip the beads out with a sieve. Place them in the drying rack.

2 **Prepare the mold.** Dry the mold out with a blast of compressed air. Spray with mold release about once each 10 times mold is used or when parts stick. Use a fluorocarbon or silicone mold release. Paste wax may also be used.

3A

Fig. 4-44. Dry filling a mold with expanded polystyrene beads.

3A (a) **Dry Fill Method:** Close the mold. Bolt or clamp shut. Open the filler hole and pour in dry pre-expanded beads. Shake down until full. Close filler hole.

3B

Fig. 4-45. Wet fill of polystyrene beads in a mold.

3B (c) **Wet Fill Method:** Fill each half of the mold with wet beads.

3C

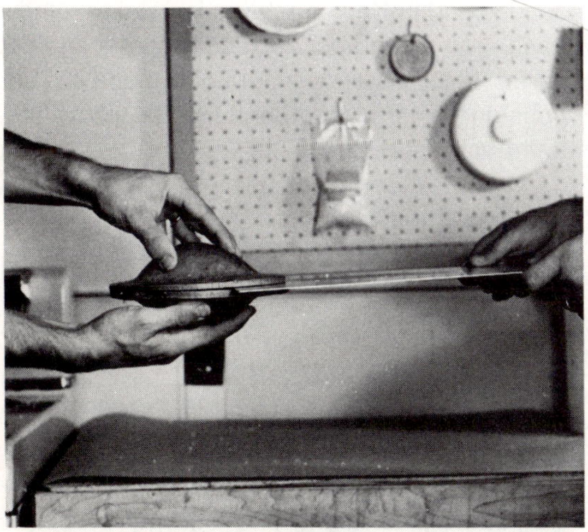

Fig. 4-46. Pulling the slip sheet.

3C Place a piece of sheet metal over one half of the mold. Put the two mold halves together. Slide out the sheet metal.

STUDENT ACTIVITIES

1. Process available resin and bead foams in school-made or purchased molds.
2. **Materials experimentation.**
   a. Pre-expand foam beads to different densities. Mold products from the different density beads. Compare them.
   b. Mix different colors of foam beads to obtain color variations.

 4

Fig. 4-47. Fusing bead foam in a pressure cooker.

4 (a) **Fusing the Foam. Pressure Cooker Method:** Bring 1 to 2 inches (25 mm - 50 mm) of water to a boil in a pressure cooker. Load several molds in the cooker. Close the cover and release valve. Bring to about 20 psi (138 kPa). Hold this pressure for 20 to 30 seconds. Remove the cooker from the stove and open the release valve. Let out **all the steam. Then** open the lid.
Caution: Steam is hot. It burns. Handle carefully.

(b) **Fusing in Boiling Water:**
Molds may be placed in pail 3/4 full of boiling water to fuse the beads. A longer heating cycle is needed. Twenty to 30 minutes is often needed for this method.

5 **Cool the mold. Put on gloves.** Take the molds from the pressure cooker or water pail. **Use a pair of pliers to pick up the molds.** Drop the molds in a bucket of cool water. Let them cool — the longer the better.

6 **Remove the part from the mold.** When you can hold the mold in your bare hands, open it. Remove the part with a blast of compressed air. Trim off any flash with a scissors.

3. **Process experimentation.** Process foam beads in the mold for different time periods. Compare the results for over- and under-processed parts.
4. Design products and make molds for resin and bead foam processes. Cast or machine aluminum molds for bead foam. Cast resin foam molds from urethane or silicone rubber. Resin foam molds may also be cast from plaster and aluminum.

SECONDARY PROCESSES

Chapter 5

There are a number of secondary processes. These processes include fabrication (construction), finishing, bonding, fastening, and decorating. Some coating operations are also considered finishing operations. Some of the secondary or finishing processes are described in this chapter. Various methods are also explained. Some are similar to those used for woods and metals.

FABRICATION

Machining Plastics

Machining may be a processing operation. Machine tools are used. Many standard sizes of plastic sheets, rods, and tubes are machined. These are used in the construction of plastic parts and products. Commercially, few of these parts or products are machined.

Machining is also a finishing operation. It follows the molding process. Many molded items require that sprue parts or flash be removed. This may be done by sawing, shearing, die cutting, and heated tool cutting. Holes are sometimes drilled in parts rather than molded in. Parting lines or trim lines may be filed and sanded in some operations. Scraping is used to remove flash on soft, low-melting temperature thermoplastics. Sanding and buffing are often used after other machining operations. This is done mostly on harder plastic materials. Acrylics and thermosetting plastics may be sanded and buffed. Tumbling is used to remove flash from compression and transfer molded parts. It is more economical. Sanding and polishing a large number of parts would take too much time.

Cutting Plastics

Plastic materials are often cut on woodworking and metalworking machinery.

Safety Note

Safety glasses should always be worn during all power cutting operations.

Sawing. Straight line cutting of plastic sheet stock is done on circular and radial arm saws. See Fig. 5-1. Fine-tooth (12 teeth per inch or 25 mm), hollow-ground, or slightly set circular saw blades work best for plastic sheet, rods, and tubes.

A variety of straight lines and curves may be cut on the band saw. See Fig. 5-2. Shark-tooth wood/metal band saw blades with four to six teeth per inch (25 mm) work best on most plastics.

Some materials require the use of blades tipped with carbide. These blades should be used to cut high pressure laminates that are reinforced with glass. Examples of materials are Formica®, Micarta®, etc.

Fine lines and sharp curves may be cut on the jigsaw. Skip tooth blades or blades with good set are necessary.

Handsaws, backsaws, and coping saws may be used for hand cutting of thermoplastics. Fine-toothed blades work best. Thermosetting plastics should be cut with a hacksaw if done by hand. This is especially important for high pressure and fiberglass reinforced laminates.

Fig. 5-1. Cutting on the circular saw. (Guard removed for illustrative purposes.)

When large holes are to be cut in plastic, a hole saw should be used. The hole saw should be fed into the plastic slowly. Use a slow spindle speed on the drill press or other driving unit.

Routing. Routers are ideal for cutting thick thermoplastic sheet products. The router is usually mounted cutter up in a table. See Fig. 5-3. A template, or guide for cutting, is used. It is made of 1/4″ (6 mm) double-tempered hardboard (Masonite® or similar material). The plastic part is placed over the template and jig. They are then guided around the template guide on the router base until the part is cut out.

Shearing. Most thermoplastic sheet stock can be cut by shearing. Shearing is safer. It makes a smoother cut. Brittle plastics like acrylic are hard to shear.

Heavy duty scissors or tin snips may be used. Thicknesses up to about 1/10″ or 3 mm may be sheared this way. A paper cutter works well for thermoplastic sheet and film up to about .030″ or 1 mm. A sheet metal squaring shear should be used for thicker sheet stock. See Fig. 5-4.

Die Cutting. Thermoformed and other plastic parts are often trimmed or cut out by die cutting. Two general types of die cutting are used:

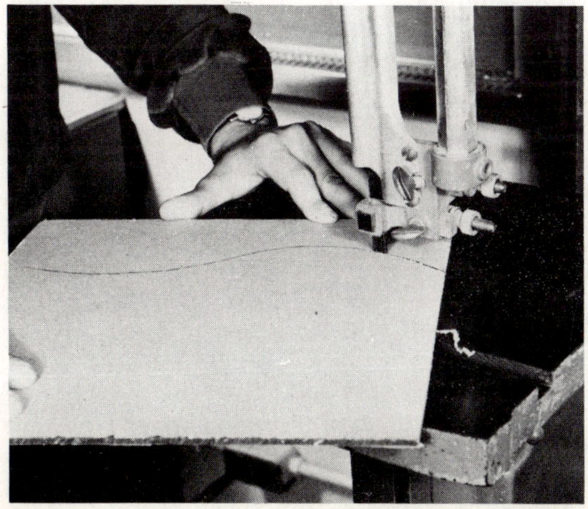

Fig. 5-2. Cutting plastic sheet on the band saw.

Fig. 5-3. Using a router to trim a part. Part is placed over the template for the cutting procedure.

cold dies and **hot dies.** Cold dies are used on plastic up to about .040″ or 1 mm thick. Hot die cutting is usually required for plastic over .040″ or 1 mm thick. Cold dies are forced through the plastic. Either a rapid stroke or high pressure is used.

Hot die cutting uses a heated die. It melts its way through the plastic. Much lower pressures are required to force the die through the plastic.

Di Arco Div., Houdaille Ind., Inc.

Fig. 5-4. Squaring shear cutting sheet plastic.

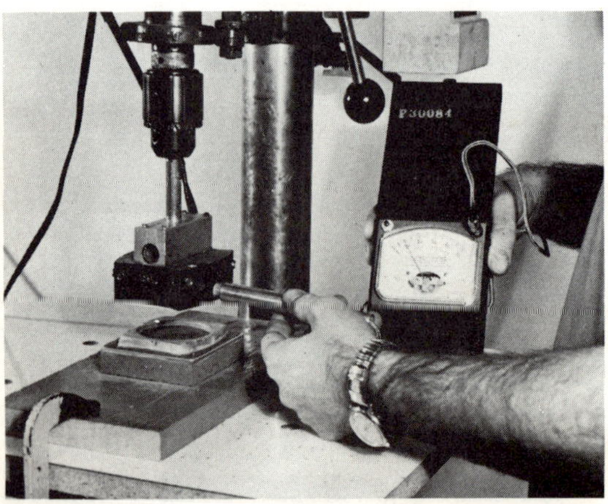

Fig. 5-5. Hot die cutter and cutting jig mounted on a drill press. Temperature is checked with a pyrometer.

A hot die cutter may be fastened to a drill press. See Fig. 5-5. Or it may be on the upper platen of a thermoformer, or on a compression press. It may also be fastened to a die cutting press. The die block and hot cutting blades should be well insulated from the press. This will prevent too much heat loss.

Hot Wire Cutting. Styrofoam and expanded polystyrene bead foam may be cut with a hot wire cutter. These low cost cutters are available in hobby craft stores.

Drilling

Most plastics may be drilled with metalworking drill bits. Brittle plastics often need blunter point angles. Smaller relief angles are also required. Metalworking bits in fractional sizes should be used. They should be sharpened with a flat lip and low lip relief angle. Drill speeds vary in the same proportion as in drilling metal.

FINISHING

Filing

Filing will remove flash or scrap from plastic parts. Large amounts can be removed in this manner. Filing works best on harder plastics. It does not generate much heat. File selection is the same for plastics as for wood and metal. Files need to be cleaned often during use on plastics.

Safety Note

Files should always have handles. This will prevent an injury. The tang of a file will cut the palm of the hand easily if a handle is not used.

Scraping

Hand scrapers and knives are often useful for removing flash or trimming edges. They are used on soft thermoplastics. Sharpened scrapers are pulled across the work. Many times they eliminate the need for sanding.

Sanding

Usually, plastic parts are sanded. This is done after excess materials are removed by cutting, filing, or scraping.

Machine Sanding. Harder plastics can be machine sanded. Coarse grit (60 to 100) open-coat sandpaper should be used. Hand sanding is usually necessary after belt or disc sanding. Fine machine sanding may be done on hard plastics. A high-speed portable orbital sander is used.

Hand Sanding. Sanding should be done in stages. Coarse garnet or production sandpaper is used first. Sanding direction is changed 90° each time a finer grit is used. Final sanding should be done with 400 or 600 grit wet-or-dry sandpaper with water. The plastic will then be ready for buffing.

Buffing

Buffing is done with a rotating cloth or flannel wheel and fine abrasive. Most plastics may be buffed. Machine buffing causes frictional heat. This softens some plastics, like polyethylene. It prevents them from being buffed.

A 6″ to 8″ (150 mm to 200 mm) loose-sewn cotton or flannel buffing wheel is best for buffing plastics. It should be driven at 1725 rpm or less.

A buffing compound is used. Red rouge or tripoli is suitable for rough, fast buffing. It is applied by holding the bar against the turning wheel. The wheel should rotate toward the operator. See Fig. 5-6. The part is moved around as it is buffed. Plenty of buffing compound should be kept on the wheel.

A final shine can be put on the part with white diamond or a similar finishing compound. The same procedures are used. A separate buffing wheel is required for each buffing compound.

Solvent Polishes

Many plastics may be polished by solvents. This is usually done only on the edges or in holes.

Solvent polishing of acrylic plastics by soaking is common in school laboratories. It is usually used for drilled holes. The hole is cleaned. It is filled with ethylene dichloride for about 30 seconds. See Fig. 5-7. The solvent is then poured out. The part is dried overnight in a well ventilated room.

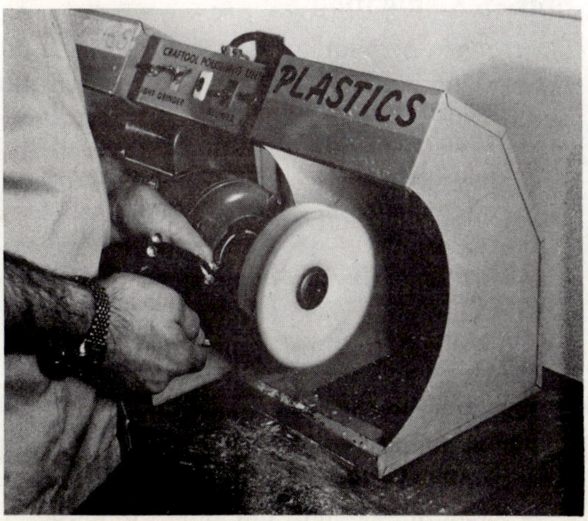

Fig. 5-6. Buffing plastics.

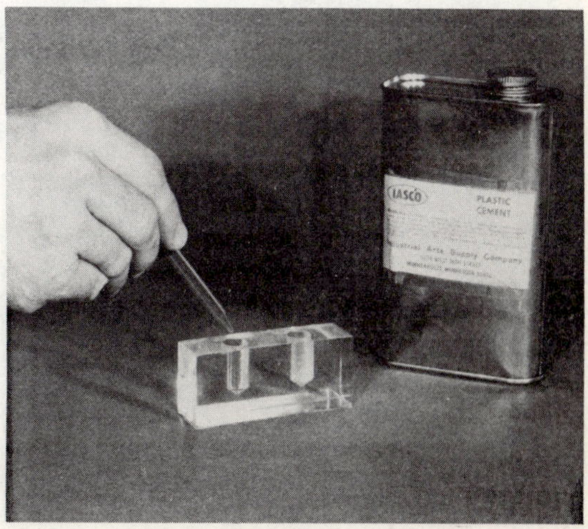

Fig. 5-7. Solvent polishing plastics.

BONDING

Often plastic parts can be joined together by cementing. The two types of cementing are adhesive and cohesive.

Adhesive Cementing

Adhesion is the joining of two plastic materials with a gluelike adhesive. There are three types of adhesive cements. These are bodied adhesives, elastomeric adhesives, and reactive adhesives.

Bodied adhesives are solvents. A small amount of the parent plastic has been dissolved in them. Styrene model airplane cement is a bodied adhesive.

Elastomeric adhesives are rubber adhesives. They are made from natural, synthetic, or reclaimed rubber. Silicone bathtub seal is an example. See how it is used in Fig. 5-8. Elastomeric adhesives will stick to almost any nonwaxy material.

Reactive adhesives are thermosetting cements. They are usually two-part adhesives. They must be mixed together just before use. An example is two-part epoxy cement.

General Electric Company

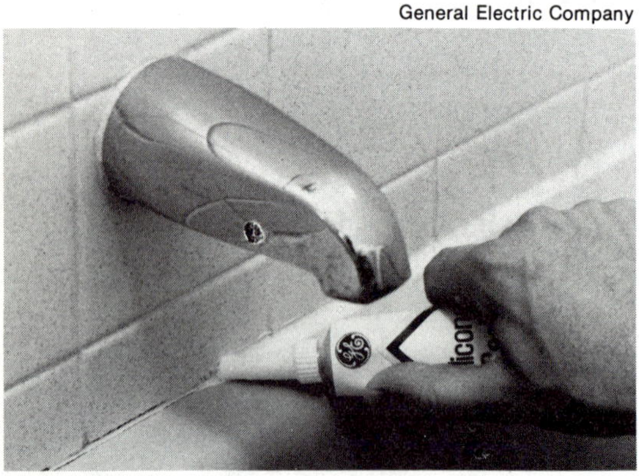

Fig. 5-8. Silicone bath tub seal.

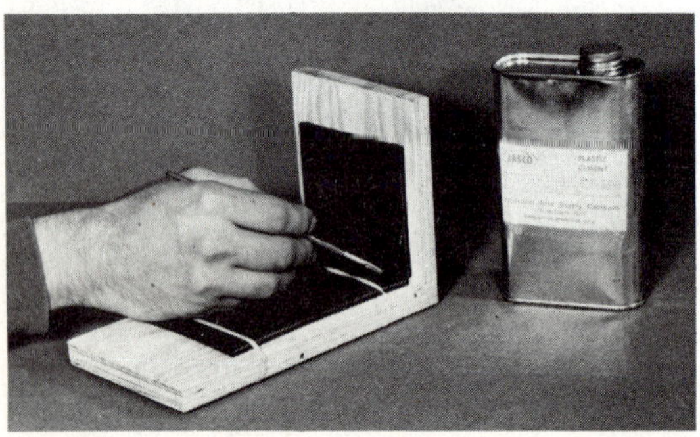

Fig. 5-9. Brush method of solvent cementing.

Cohesive Cementing

Cohesive cementing is the softening of the two plastic parts with a solvent. Once softened, the molecules in the two plastic parts mix together upon contact. The solvent evaporates. The plastic hardens and the two parts are joined as one. Thermal welding and heat sealing are also forms of cohesive bonding. They are discussed later in this chapter.

The two types of cohesive cements are solvent cements and monomeric cements.

Solvent cements will dissolve or soften the surface of the two parts to be joined. Solvent cements usually work only on thermoplastics. Several solvents are usually available for each plastic.

Monomeric cements are made from the same monomer as is the plastic to be joined. The monomer is catalyzed just before use. It cements by polymerization.

Solvent Cementing

Solvent cementing is probably most used in the laboratory. It will bond acrylic plastic parts. The building trades use it to bond ABS and PVC sewer and water pipe. Parts to be solvent-cemented should have clean joints. Follow the procedures below:

1. **Prepare the joint.** Be sure the parts fit together well.
2. **Apply the solvent cement.**
   a. **For small areas** or thin joints, the solvent may be put on with a brush, eye dropper, or hypodermic needle. See Fig. 5-9. Hold the joint together. Run a small amount along the edge of the joint. Do not get it on the sides of the plastic. Capillary at-

traction (pulling force resulting from surface tension) will draw the solvent into the joint. It may be necessary to put it on from both sides.

b. **For large areas,** soaking, dipping, or flooding may be needed.
    1) **Soaking.** Place a shallow layer (about 1/16″ or 1 mm) of solvent in a flat pan. Put the end of the part to be joined into it. See Fig. 5-10. Felt may be placed in the bottom of the pan. It will help keep the solvent from oversoaking the plastic.
    2) **Dipping.** Dip the joint into a pan of solvent for a few seconds. Take it out, and put the joint together.
    3) **Flooding.** For large areas, such as for acrylic solvent laminating, lay one piece down flat. With an eye dropper, flood part of it. See Fig. 5-11A. Put the second piece over the first. Gradually close the joint pushing the air out ahead, Fig. 5-11B.
    4) **Hold the joint together** until the solvent evaporates. Use just enough pressure to keep air out of the joint. Do not clamp too tightly. A fixture for holding the joint straight may be needed. It can be made from scrap wood. A fixture is shown in Fig. 5-9 and Fig. 5-10.

Fig. 5-10. Soak method of solvent cementing.

FASTENING

Mechanical Fastening

A variety of mechanical fastening devices may be used to hold plastic parts together securely. They are divided into two groups: add-on mechanical fasteners and built-in mechanical fasteners.

Add-on Mechanical Fasteners. Add-on or extrinsic fasteners are devices used on the outside of a product to hold the parts together. Many kinds may be used. These include screws, nuts and clips, hinges, rivets, and staples. Most of these can be purchased in hardware stores. Few require special knowledge or procedures for use.

Built-in Mechanical Fasteners. Plastic parts may be put together without external fasteners. Products may be molded with the fastening device built in. These devices are

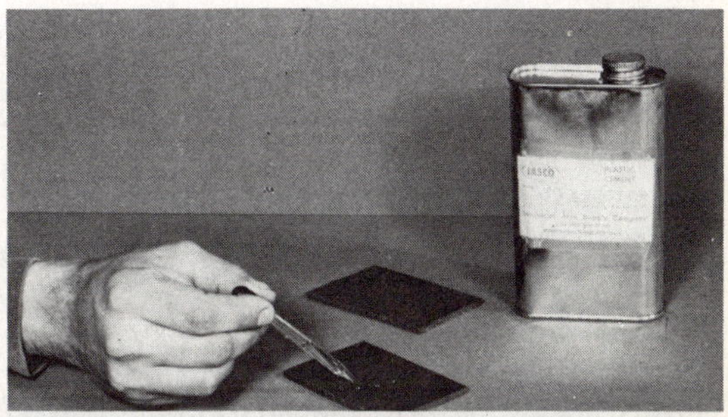

A. Use an eyedropper to flood part of piece.

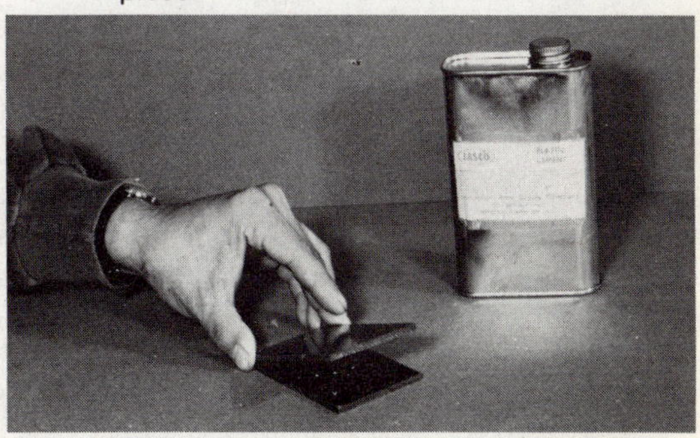

B. Place two parts together.

Fig. 5-11. Flooding method of solvent cementing

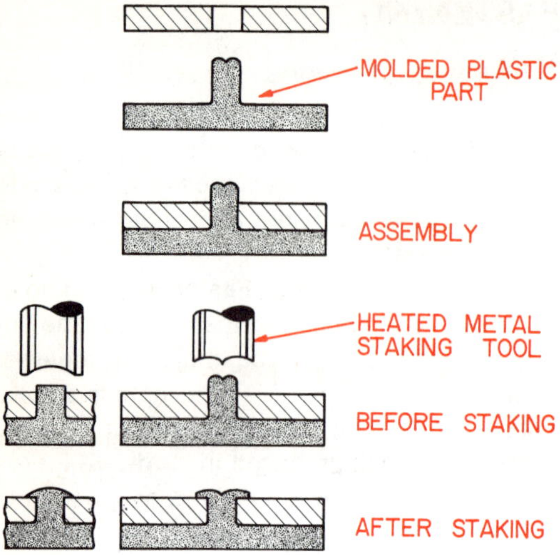

Fig. 5-12. Fastening by heat staking.

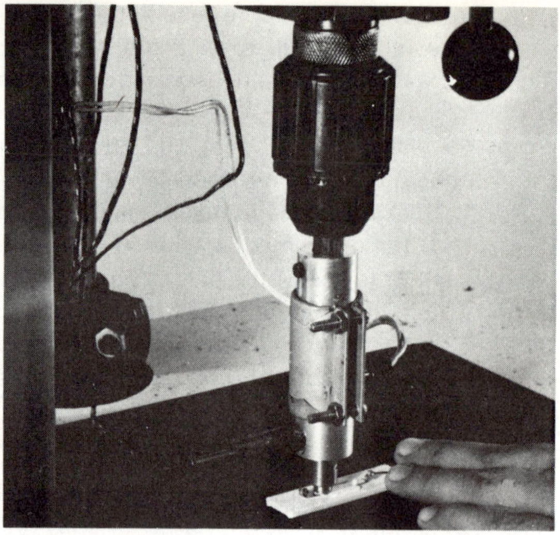

Fig. 5-13. Heat staking a pin back to a name tag.

often called **intrinsic** mechanical fasteners. The types of fasteners used are snap fit, press fit, heat staking, and ultrasonic staking.

Two types of **snap fit** fasteners may be molded into the parts. These are snap in and snap on.

In **press fit** fastening, two parts are forced together. The internal part is molded slightly too large. Such a fastener should be used only for stiff, tough plastics.

Heat staking is similar to fastening with spring clips or rivets. A plastic stud is molded into the part as shown in Fig. 5-12. It is pushed up through the second part. A heated tool is pressed down over the stud. This forms a head. When the plastic cools, the head holds the parts together.

A faster method of staking thermoplastics is **ultrasonic staking.** This method fastens parts together solidly. A vibrating tool, rather than a heated tool, forms the head. The vibrating tool is also called an ultrasonic horn. It moves up and down 20,000 times per second. The ultrasonic vibration causes frictional heat in the plastic stake under the horn. The heat forms the head on the stake.

Heat Sealing Plastics

Heat sealing is used to fasten two thin pieces of thermoplastics film or sheet together. The two layers are melted together by heat and pressure. A cohesive bond is formed.

There are four methods of heat sealing. These are thermal sealing (heated tool), impulse sealing, dielectrical sealing, and ultrasonic sealing.

Thermal Sealing. Thermal sealing is the simplest method of sealing plastic parts. It is also the least expensive and the most popular method.

Several tools may be used for thermal sealing. One is a heated tool much like a small clothes iron or soldering iron. Another is the heated wheel. Hot plates are used in meat and produce markets. Many food products are packaged by heat sealing. Some two-sided heaters are used for thicker materials. Hand-held heat sealing irons are usually inexpensive. They are easily used in the school laboratory.

Two popular heat sealing plastics are polyethylene and cellophane. Many other flexible plastic films can be succesfully heat sealed.

Following is the **thermal sealing procedure.** Use a piece of clean plastic film.

1. Plug in the sealer and warm it up.
2. Lay two pieces of plastic film together or use lay-flat plastic tubing.
3. Press the iron or roller over the plastic film. Move it at an even speed. Iron the entire length of the seal.

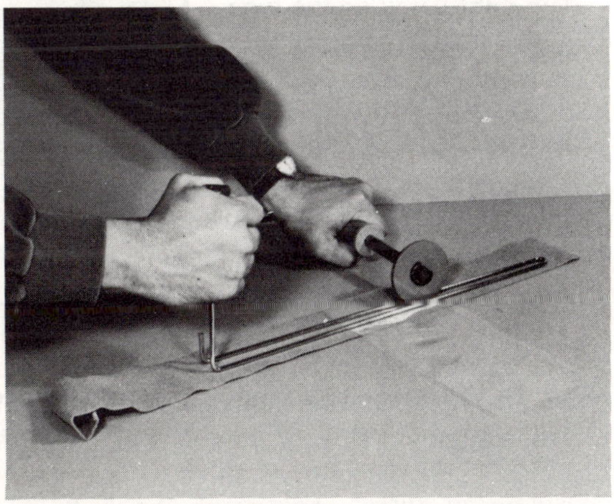

Fig. 5-14. Hand-held rotary heat sealer. Note the felt pad under the film.

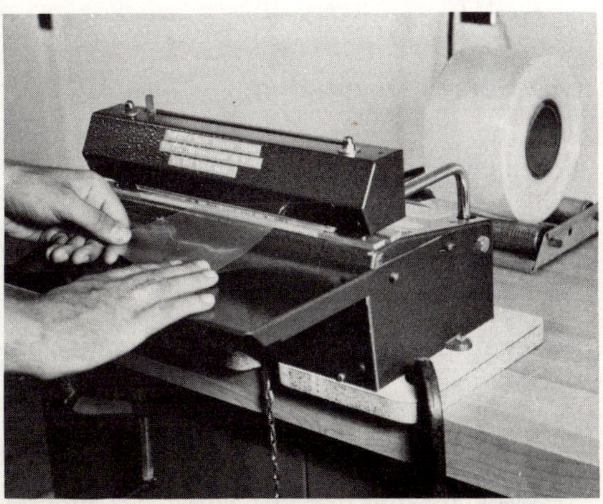

Fig. 5-15. Impulse heat sealer in use.

4. Remove the iron. Let the joint cool.

A felt pad is used under the rotary wheel heat sealer. See Fig. 5-14. It will cushion the force of the wheel. It also helps prevent cutting through.

Impulse Sealing. In impulse sealing, the plastic film is clamped together. It is heated and cooled quickly under pressure. This method is usually used for thin films. They must be .010″ (0.25 mm) or less in thickness.

STUDENT ACTIVITIES

1. Join or laminate several pieces of acrylic plastic together with solvent cement.
2. Join several pieces of ABS or PVC sewer or water pipe together with solvent cement.
3. Build plastic airplane, car, or naval models with bodied cement.
4. Use heat staking to fasten molded products together. One project might be to stake the clip on the back of a name tag.
5. Use a heat sealer (heated tool, impulse sealer, or ultrasonic sealer) to package molded parts, mass-produced products, or gifts. Make pocket savers from vinyl plastic sheet.
6. Heat-seal a quantity of water into a bag.

COATING AND DECORATING

Coating and decorating add eye appeal or beauty to a product. Plastic parts are coated or decorated for other reasons, too. These processes often improve resistance to wear. They may also prevent scratching and marring. They change the color, texture, contrast, and design of many products. Identification and product information are often placed on the product. Coating and decorating may be done either during or after the molding operation.

There are many different methods of decorating products. This chapter discusses several. Hot foil stamping, silk screen printing, and painting are presented. Plating and vacuum metalizing, electrostatic powder coating, and fluidized bed coating are also discussed.

Hot Foil Stamping

In the hot foil stamping process, a decorative coating is placed on a plastic part. It is transferred from a roll of foil or film to the part. Heat and pressure are used. Foil usually has a polyester or cellophane film backing. Heated dies or metal printing type are pressed against the uncoated side of the backing. The foil coating sticks to the plastic. It releases from the backing.

Foils come in many colors. These include real gold and silver leaf. Wood grain and other patterned foils are also used on plastics.

Names may be put on products by hot stamping. Dies for hot stamping are made from several metals. Brass, steel, zinc, and magnesium may be used. The dies shown in Fig. 5-16 are silicone rubber bonded to metal.

Laboratory Equipment and Materials

Laboratory hot foil stamping may be done with a small manual machine. Special type should be used. It should be heat and pressure resistant. Steel or brass type may be used. A more economical type is known as "service type." It is satisfactory for laboratory use.

Gladen Enterprises, Inc.

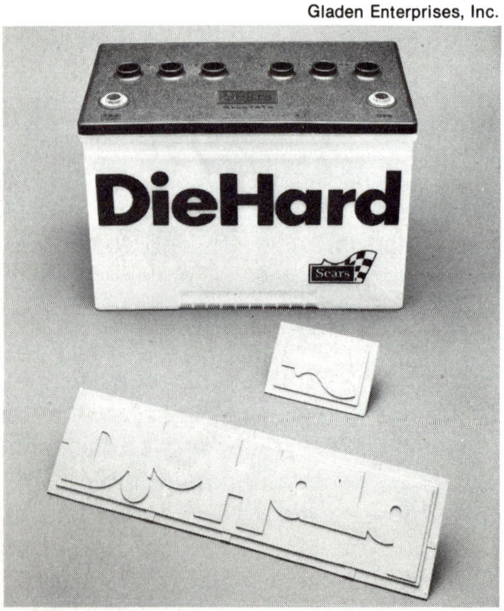

Fig. 5-16. Silicone dies for hot stamping. Stamped product shown in the background.

Operation Sequence for Hot Foil Stamping

The hot foil stamping process is done in seven steps.

1. Choose the type. Set it in the typeholder. Set it as in graphic arts. It should be upside down (nick up), face up, and spelled from left to right. Lock up the type.
2. Heat the typeholder, with the type in it, to about 225° F to 300° F (110° C to 150° C). The temperature depends upon the plastic to be stamped. Plastics with lower melting temperatures require lower stamping temperatures.
3. Choose the right foil for the plastic used.
4. Put the part to be stamped in the locating fixture.
5. Lower the die onto the plastic. Keep the stamping foil between them. Be sure the dull side (pigment side) of the foil is against the plastic. Put a slight pressure on the lever until you "feel" the plastic give slightly. **Do not over-impress the plastic.**
6. Remove the pressure. Take the foil off the plastic.
7. Remove the plastic part from the holding fixture.

Silk Screen Decorating

The silk screen process is used on plastics. It imprints both designs and line work. Silk screen printing is an inexpensive method. A durable product is produced. It may be done by hand or with automatic equipment. The production rate is 200 to 600 pieces per hour by hand. Automatically 2000 to 3000 per hour may be produced.

Painting Plastics

Most plastics may be painted. It is difficult, however, to make paint stick to waxy plastics.

Polyethylene, polypropylene, and fluorocarbon do not take paint easily. Plastics are usually painted for decoration. Colors contrasting the molded-in color of the plastic part are used most often. Both enamel and lacquer base paints are used. The type of paint depends on the plastic. Several painting methods may be used. Some of these are:

1. Spray painting the entire surface.
2. Mask spray painting.
3. Flow coating.
4. Dip coating.
5. Roller coating.
6. Spray and wipe painting.

Plating and Vacuum Metalizing

A bright metal coating may be applied to plastic parts. It is put on either the front (first surface) or back (second surface) of the part.

Dependable Machine Co., Inc.

Fig. 5-17. Products printed by the silk screen process.

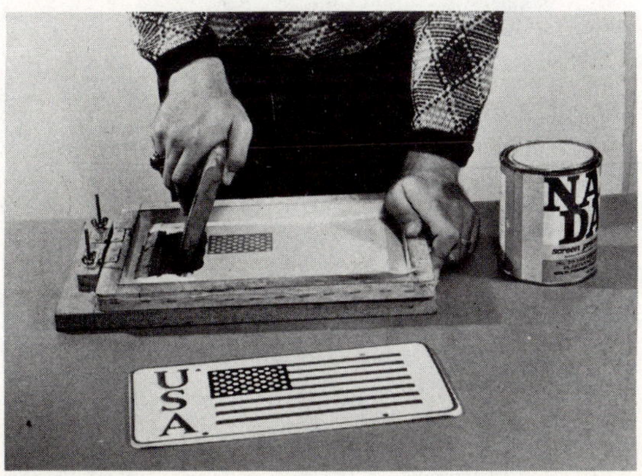

Fig. 5-18. Silk screen printing in the laboratory.

Two types of bright metal coatings are generally used. These are:

1. Electroplating and
2. Vacuum metalizing.

Of the two, electroplating is the most durable. Vacuum metalizing is less expensive. Plating is usually done only on the first surface. Vacuum metalizing may be done on either surface.

Electroplating. Plastics may be electroplated. First, an electrically conductive coating of copper must be chemically bonded to the cleaned plastic surface. Then metals are electrically placed on the copper base coat. Bright chrome, silver, nickel, or gold may be used.

Electroplating provides a sturdy, bright finish. It will protect the plastic part. The light weight of the plastic and the bright metal finish produce unusual items. Many automotive and plumbing industry products are electroplated. See Fig. 5 10.

Vacuum Metalizing. Vacuum metalizing is used for plating items such as plastic model kit parts, decorative automobile parts, and household items. Some are shown in Fig. 5-20. The plastic is first cleaned. It is then sprayed with a

Div. General Motors Corp.

Fig. 5-19. Inspecting an electroplated automobile grill.

base coat of lacquer. The dried base coat later bonds the metallic coat to the plastic. The parts are put on rotating racks. The center of the rack has tungsten wire filaments. Small metal clips, usually aluminum, are on the filaments. The filled rack is pushed inside a large steel vacuum chamber. The door is closed. A near-perfect vacuum is pulled on the closed chamber. The tungsten filaments are then electrically heated to about 1800° F (1000° C). The metal on the filament "flashes" or evaporates from the heat. The flashed metal travels in a straight line. It sticks to any surface it hits. The parts receive a coating.

Premier Vacuum Process

Fig. 5-20. Vacuum metalized products.

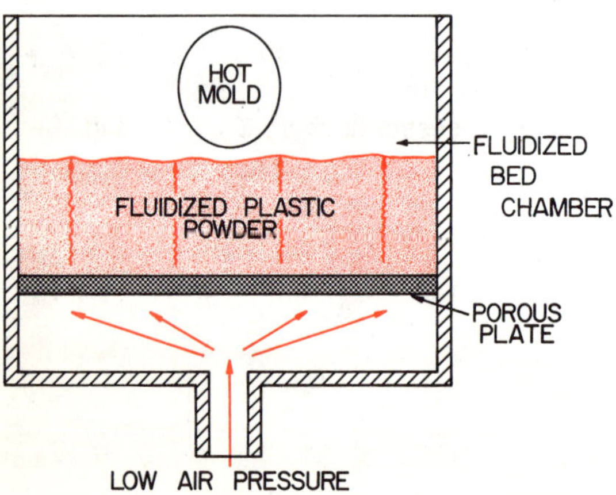

Fig. 5-21. Fluidized bed coating.

Fluidized Bed Coating

Fluidized bed coating is a process that uses dry powdered plastic materials. Polyethylene is very popular for this process. The dry plastic powder is put in a chamber called a **fluidized bed.** See Fig. 5-21. Air is blown up through a screenlike porous plate in the bottom of the chamber. The moving air sets the dry powdered plastic in motion. The dry powder tends to float on the air. It acts like a fluid (liquid). This is why it is called "fluidized."

An object (or mold) to be coated is first heated. It is then lowered into the fluidized powder. The powder melts and fuses to the surface of the object. A layer of plastic builds up. The object is taken from the powder and put back in the oven. The heat smooths the plastic

Polymer Corp.

Fig. 5-22. Fluidized bed-coated lawn furniture in use at Knott's Berry Farm. The plastic coating is more durable than paint.

surface. The object is then taken from the oven and cooled.

Checklist for Materials and Equipment Needed

Assemble all necessary materials for fluidized bed coating.

1. Oven, electric stove, or rotational molder oven.
2. Fluidized bed coating machine.
3. Powdered plastic. Coating grade polyethylene works very well. A molding grade may be substituted.
4. Primer — if needed.
5. Part to be coated.
6. Short piece of wire.
7. Insulated gloves and long-nosed pliers.
8. Dry color pigments — optional.

1

Fig. 5-23. Hang part on a wire in oven to heat.

1 Hang the part on a wire. Bend the other end of the wire to form a hook. Hang the part in the oven. Preheat it to about 100° F (38° C) above the melting point of the plastic being used. About 10 to 15 minutes at 350° F (177° C) for polyethylene will work.

 2

Fig. 5-24. Dip in fluidized bed.

2 (a) Fill the fluidizer about 1/2 to 2/3 full of plastic powder. Add a dry color pigment **if desired.**

(b) Turn on enough air pressure to **float** the plastic powder. About 3 to 5 psi (20.7 to 34.5 kPa) is enough.

3 **Put on insulated gloves.** Take the part or tool from the oven with a pair of pliers. Dip it in the fluidized plastic **immediately.**

4 When the part stops picking up plastic powder, put it back in the oven. Fuse it for about 2 to 4 minutes.

5 Extra coats of plastic may be added by the same method.

STUDENT ACTIVITIES

1. Use the hot foil stamping process on plastic products. The products may be made by such processes as thermoforming and injection, rotational, and blow molding.
2. Set up a silk screen for decorating thermoformed and blow-molded products.
3. Build a fluidized bed coating machine. Coat products.
   a. Experiment with different grades of powdered polyethylene. Decide which gives the best coating and sticks the best.
   b. Color some polyethylene powder for fluidized bed coating.

CAREERS IN THE PLASTICS INDUSTRY

Chapter 6

The plastics industry offers a variety of career opportunities. The amount and type of education and experience needed depends upon the job. Many jobs are available to people with limited education and experience. However, those who have special training or experience can generally expect better salaries. In this chapter, several types of jobs are discussed. Consider the qualifications you would need to be hired. Consider what your duties would be. Think about each job as you read. Where might you fit into the plastics industry?

Finishers and Assemblers

Plastic parts and products usually require hand or machine finishing after molding. Many require assembly.

Finishing may be done by the machine operator. If too much finishing is required, the parts may be sent to a **finishing department. Finishers** will perform these operations. They may trim, scrape, sand, buff, or drill holes in the parts.

A large plant may have an **assembly department.** Here the parts are put together. Brand

Williams - White & Company

Fig. 6-1. Structural foam molding press machine operator.

National Assoc. of Women in Construction

Fig. 6-2. Design engineering. One of the first steps in the development of a new product idea is creating its design.

Beloit Corp.

Fig. 6-3. Inspection is an assembly function. Final products are checked. Here, quality control supervisors examine polyethylene containers.

Portage Industries Corp.

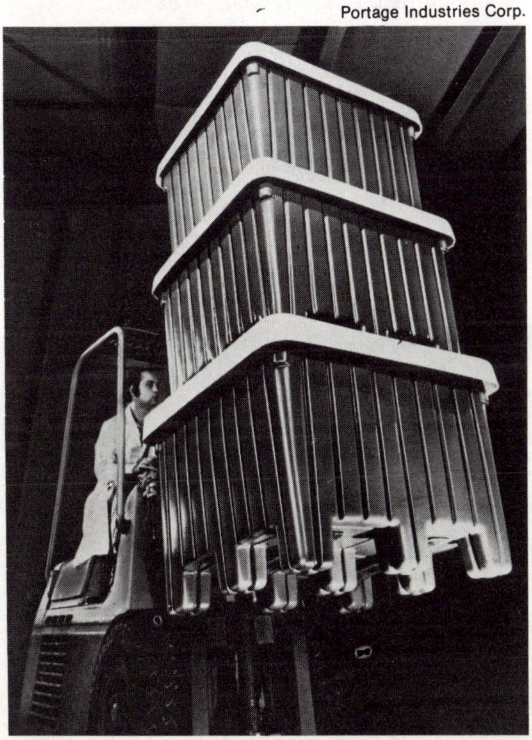

Fig. 6-4. Workers use machines to carry materials and finished products.

name labels or finishes are also put on. Typical **assembler** duties may include putting on handles, knobs, or labels. Plastic parts may be assembled by spin welding, cementing, ultrasonic welding, staking, or with metal fasteners. Finished parts or assemblies are often packed for shipment on the assembly line.

New plant employees are often assigned to the finishing and assembling departments. No previous experience or training is necessary. People who apply for the job should be dependable. They must be willing to repeat the same tasks many times. Wages are similar to assemblers and finishers in other industries in the same geographic area.

Floor Workers and Material Handlers

In large plants, molding machines may run continuously. In such cases, three shifts of workers are used. Materials must be continually supplied to the molding machines. People move the materials. These people are called **material handlers** or **floor workers.** Lifting and carrying are often necessary. A material handler may have to move 50-pound (25 kg) bags of plastic by hand. Fork lift trucks and other mechanical handling equipment are also used. Some larger plants are installing bulk handling systems. These systems eliminate some handling jobs.

The job qualifications and wages of material handlers and floor workers are generally the same as assemblers and finishers. New people are usually assigned. No experience is necessary.

Machine Operators

Molding and forming machines are used in the plastics industry. These require **operators.** Often a machine operator will run more than one machine. The operator may also perform some finishing tasks. This is more likely if only one machine is being operated.

Machine operators learn their jobs by doing them. They learn to operate the production

Frye, Inc.

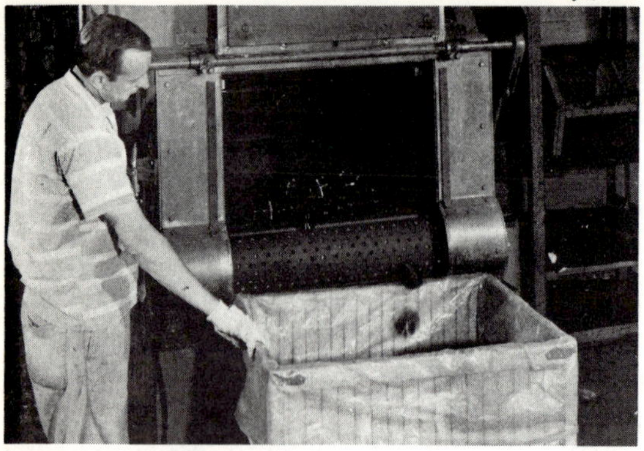

Fig. 6-5. Operating a commercial tumbler in the finishing department.

machines in the plant. Machine operators make only a certain part. When a new part is set up, the operator will learn new duties. New machine operators are usually assigned to a machine already in operation.

Machine operators are responsible for the quality of their molding machine product. This usually requires visual inspection. They look for rough edges, flaws, or defects. Operators

U.S. Industrial Chemcials, Co.

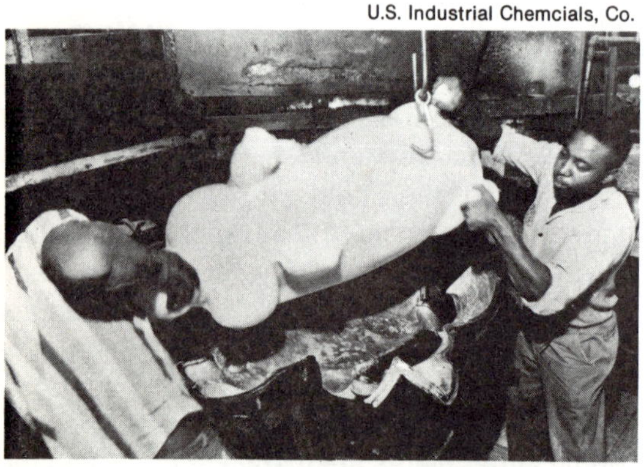

Fig. 6-6. Machine operator removing a large, rotationally molded part from the mold.

Beloit Corp.

Fig. 6-7. Machine operator running a blow molder.

Beloit Corp.

Fig. 6-8. Machine operator.

seldom adjust their machine. They will call for help if this is needed. The adjustments are usually assigned to supervisors, trouble shooters, or cycle analysts.

Machine operator is also a beginning job. Wages are similar to those paid assemblers, finishers, and floor workers.

Mold Setters and Maintenance Mechanics

Machine adjustments are usually made by persons called **set-up workers, cycle analysts, maintenance workers,** or **trouble shooters.** These job titles may be different from one plant to another. Job functions vary. People in these jobs work with a variety of hand and power tools. They must understand the molding process. They must also know and understand their assigned machines. They will set up, adjust, and maintain the machines and molds. This group of workers must be good plumbers, electricians, and mechanics.

Set-up workers are responsible for setting up the machine. They must adjust it to operate properly. They must understand and follow directions prepared by a production engineer for cycle times, temperatures, and pressures.

Set-up workers, cycle analysts, maintenance workers, and trouble shooters are often experienced machine operators. They are trained on-the-job for these new assignments. Some trade schools, vocational schools, and community colleges also provide training programs. Wages are similar to those paid maintenance people in other industries. Generally they are good paying jobs.

Tool and Die Maker

Many plants employ tool and die makers. They are called **mold makers** in the plastics in-

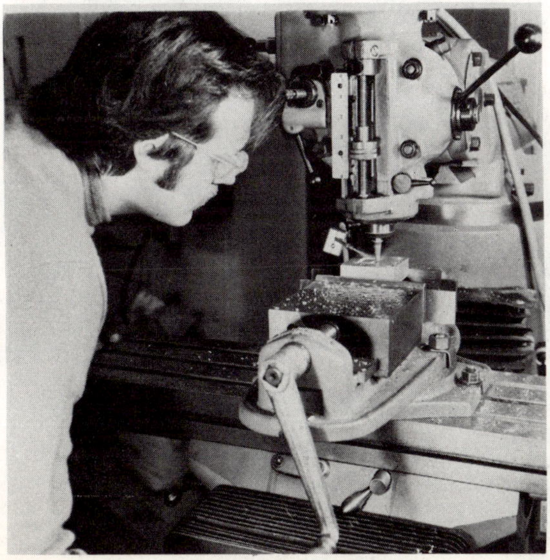

Fig. 6-9. Tool and die maker.

Beloit Corp.

Fig. 6-10. Inspecting an injection molded part.

dustry. They are skilled metalworkers. Mold makers may not work in the plastics plant. They may be employed by shops who make molds for the plastics industry. Mold makers work with equipment that is very exact. They use milling machines, lathes, and surface grinders. Mold makers build or repair molds. Since accuracy and skill is required, these people require special training. They may receive it on-the-job as apprentices. Vocational or technical schools may offer training programs.

Wages for tool and die or mold makers are excellent. These people are in constant demand. There is usually a shortage of them.

Production Engineer

Production engineers design the molds. They also make plans for the molding of a part or product. They must work closely with the designers of a product. The designers are called **design engineers.** Production engineers also work closely with the people who set up and operate the process.

Production engineers usually have a college degree in engineering. It is often mechanical or chemical engineering. Salaries and fringe benefits are good. They are similar to those given people with engineering degrees in other industries.

Fig. 6-11. Production engineers plan the molding of a product.

APPENDIX

TABLE 1

CHRONOLOGY OF PLASTICS*

DATE	MATERIAL	EXAMPLE
	CIVIL WAR	
1868	Cellulose Nitrate	Eyeglass Frames
1909	Phenol-Formaldehyde	Telephone Handset
1909	Cold Molded	Knobs and Handles
	WORLD WAR I	
1919	Casein	Knitting Needles
1926	Alkyd	Electrical Bases
1926	Analine-Formaldehyde	Terminal Boards
1927	Cellulose Acetate	Toothbrushes, Packaging
1927	Polyvinyl Chloride	Raincoats
1929	Urea-Formaldehyde	Lighting Fixtures
1935	Ethyl Cellulose	Flashlight Cases
1936	Acrylic	Brush Backs, Displays
1936	Polyvinyl Acetate	Flash Bulb Lining
1938	Cellulose Acetate Butyrate	Irrigation Pipe
1938	Polystyrene or Styrene	Kitchen Housewares
1938	Nylon (Polyamide)	Gears
1938	Polyvinyl Acetal	Safety Glass Interlayer
1939	Polyvinylidene Chloride	Auto Seat Covers
1939	Melamine-Formaldehyde	Tableware
	WORLD WAR II	
1942	Polyester	Boat Hulls
1942	Polyethylene	Squeezable Bottles
1943	Fluorocarbon	Industrial Gaskets
1943	Silicone	Motor Insulation
1945	Cellulose Propionate	Automatic Pens and Pencils
1947	Epoxy	Tools and Jigs
1948	Acrylonitrile-Butadiene-Styrene	Luggage
1949	Ailylic	Electrical Connectors
	KOREAN WAR	
1954	Polyurethane or Urethane	Foam Cushions
1956	Acetal	Automotive Parts
1957	Polypropylene	Safety Helmets
1957	Polycarbonate	Appliance Parts
1959	Chlorinated Polyether	Valves and Fittings
1962	Phenoxy	Bottles
1962	Polyallomer	Typewriter Cases
1964	Ionomer	Skin Packages
1964	Polyphenylene Oxide	Battery Cases
1964	Polyimide	Bearings
1964	Ethylene-Vinyl Acetate	Heavy Gauge Flexible Sheeting
	VIETNAM WAR	
1965	Parylene	Insulating Coatings
1965	Polysultone	Electrical/Electronic Parts
1966	Polyphenyl Oxide (PPO)	Electrical/Electronic and Appliance Parts
1970	Thermoplastic Polyester	Electrical/Electronic Parts
1973	Polybutylene	Piping
1975	Nitrile Barrier Resins	Bottles

*From the booklet ***The Story of the Plastics Industry*** (SPI)

Table 2

Resin	Melting Point °F	(°C)	Specific Gravity[1] @ 23° C.	Examples of Materials
Acrylic - rubber modified	279	(137)	1.12 - 1.16	"IMPLEX"[2]
Acrylonitrile - Butadiene - Styrene	325	(163)	1.04	"CYCOLAC"[3]
Acrylonitrile - Styrene	268	(131)	1.08 - 1.06	"TYRIL"[5] "FOSTACRYL"[6] "LUSTRAN A"[18]
Cellulose Acetate	[1]446	(230)	1.25 - 1.35	
Cellulose Acetate Butyrate	[1]356	(180)	1.17 - 1.24	"TENITE" BUTYRATE[9]
Cellulose Nitrate	Decomposes		1.34 - 1.40	"XYLONITE"[10]
Cellulose Propionate	[1]456	(236)	1.20 - 1.24	"FORTIFLEX"[11]
Cellulose Triacetate	572	(300)	1.27 - 1.28	"B X TRIACETATE"[10]
Chlorinated Ether	358	(181)	1.4	"PENTON"[8]
Ethyl Cellulose	—		1.10 - 1.16	"ETHOCEL"[5]
Fluorocarbons				
FEP	554	(290)	2.10 - 2.19	TEFLON® [1]
TFE	621	(327)	2.14 - 2.17	TEFLON[1]
Fluorochlorocarbons	383	(195)	2.1	"KEL-F"[13]
Fluorohydrocarbons				
Vinylidene Fluoride Polymers	340	(171)	1.76	"KYNAR"[17]
Methylstyrene	349	(176)	1.03	"CYMAC" 400[4]
Nylon				
66 nylon ZYTEL® 101	489	(254)	1.14 - 1.15	ZYTEL® [7]
ZYTEL 42	489	(254)	1.14	
610 nylon ZYTEL 31	415	(213)	1.09	
6 nylon ZYTEL 211	405	(207)	1.13	
11 nylon 'RILSAN'	351	(177)	1.04	
alcohol soluble				
nylon: ZYTEL 63	311	(155)	1.08	
ZYTEL 69	293	(145)	1.08	
Polyacetal	347	(175)	1.42 - 1.43	DELRIN® [1]
Polyacrylate & Methacrylate	374*	(190)	1.18 - 1.19	LUCITE® [1]
Polycarbonate	430	(221)	1.2	"LEXAN"[7]
Polyester Films	482	(250)	1.38 - 1.40	MYLAR® [1]
Polyethylene	221	(105)		ALATHON® [1]
	low density		.914 - .930	
	248	(120)	.94 - .96	
	high density			
Polypropylene	334	(168)	.85 - .91	ALATHON® [1]
Polystyrene	374	(190)	1.04 - 1.08	"DYLENE"[12]
Polystyrene - Methylmethacrylate	351	(177)	1.14	"ZERLON"[5]
Polyvinyl Acetate	140 to 190	(60 to 88)	1.18 - 1.20	ELVACET® [1]
Polyvinyl Alcohol	446	(230)	1.27 - 1.31	ELVANOL® [1]
	Decomposes			
Polyvinyl Butyral	[2]345	(174)	1.07 - 1.20	BUTACITE® [1]
Polyvinyl Chloride Polymers[3]	302	(150)	1.30 - 1.60	"EXON"[16] "VIPLA"[14]
Polyvinyl Chloride - Vinyl Acetate	261	(127)	1.34 - 1.37[2]	"VELON"[16]
Vinylidene Chloride Polymers	313	(156)	1.63 - 1.73	"SARAN"[5]

1 E. I. du Pont de Nemours & Co. (Inc.)
2 Rohm and Haas Co.
3 Marbon Chemical Div., Borg-Warner Corp
4 American Cyanamid Co.
5 The Dow Chemical Co.
6 Foster Grant Co.
7 General Electric Co.
8 Hercules Powder Co.
9 Eastman Chemical Products, Inc.
10 BX Plastics Ltd.
11 Celanese Plastics Co.
12 Koppers Co.
13 Minnesota Mining and Manufacturing Co.
14 Chemore Corp.
15 Dayton Industrial Products Co.
16 Firestone Plastics Co.
17 Pennsalt Chemical Co.
18 Monsanto Company

***Will vary between grades 6, 7, or 8**

1 Specific gravity is the relationship of the weight of any material to the weight of pure water.

2 Wide range of melting points, depending upon composition.

3 Plasticizers will usually lower specific gravity and/or may mask odor.

INDEX